HYDRAULIC AND PLACER

MINING

BY

EUGENE B. WILSON

SECOND EDITION, REWRITTEN

FIRST THOUSAND

NEW YORK

JOHN WILEY & SONS

LONDON: CHAPMAN & HALL, LIMITED

1907

PREFACE TO SECOND EDITION.

THE demand for the first edition of this work, and the great activity developed in placer mining, due in a large measure to the great returns from this species of work, as well as the very substantial profit accruing to the exploitation of the placers, has led the author to present this second edition.

There have also been many new methods for catching the free gold, as well as great improvements in the machinery for handling the material, and in the application of new machinery to placers where unusual difficulties were encountered in working them.

All these considerations have led the author to issue the new edition, which in his opinion, brings this work abreast of the latest improvements in this industry. He desires to acknowledge his indebtedness to the various technical journals, as well as to the engineers whose names appear in the text.

The manner in which this subject is presented, the author thinks, will appeal not only to those engaged in placer mining, but to those who desire to get the latest ideas relating to this industry.

THE AUTHOR.

CONTENTS

CONTENTS

HYDRAULIC AND PLACER MINING

CHAPTER I.

GEOLOGY OF PLACER DEPOSITS.

THE term placer is defined as a place where surface depositions are washed for the valuable minerals, gold, tin, tungsten gems, etc. Placer mining is defined as washing surface depositions for gold.

The gold found in alluvial deposits is in the metallic state, and in all probability was derived from the disintegration of gold-bearing rocks and veins. The chemical and mechanical processes that freed the gold from the rocks with which it was associated are numerous and sometimes obscure, although it is obvious that the elements in combination with water and ice have been the chief factors in tearing rocks apart and concentrating the gold.

Most placers are composed of quartz pebbles, sand, rocks of various descriptions and sizes, and clay. The gold forms a very small portion of the mass, while quartz sand, with garnets and black sand, the latter composed of magnetite, ilmenite, and hornblende, make up another portion of the mass.

The disintegration of rocks is occurring continually, and has been for ages. The lighter particles are carried away by wind and water, and the heavier are left. While

the so-called elements — wind, heat, cold and water —
have been the chief causes of disintegration, there are
indications that glaciers have played an important part
in the erosion of rocks, as they moved from higher to
lower levels, particularly in the more northern latitudes.
Water has been the chief factor in the formation and
concentration of placers, although the other elements
have assisted.

The San Juan River extends from North Bloomfield
to Nevada City, California. This was an ancient and
by some called a tertiary river, that cut a natural channel through the bed rock. In time the channel became
partially filled with wash dirt and gravel that contained
gold, until the deposits reached a thickness of 500 feet.
When the upheaval occurred that formed the Sierra
Mountains the river bed was raised high above its original position. Lava flows later followed the upheaval,
filled the ancient river channel, and covered the placer
in the river bottom to a depth of from 200 to 400
feet. As the width of this river varies from one to
one and a half miles, it must have been an important
water course in its time. The present river follows
the course of the ancient river; however, as this is but
one of many ancient rivers it is reasonable to expect
that some of the modern rivers would cut through the
lava capping of the old placers to bed rock and
form modern placers. This is the case, so that two
kinds of mining are needed to extract the gold from the
two differently situated ancient placers. Where the
ancient river beds have been eroded by modern rivers
cutting across the overlying capping of lava, secondary

placers have been formed which also are a source of gold.

The composition of the gravels in these ancient river channels represents nearly every rock in the vicinity, such as diabase, diorite, serpentine, slate, granite, syenite, and quartz. The lowest gravels in some placers have a blue tint, and are termed the "blue lead" since in this locality the richest ground is found above bed-rock. The depth of the "pay streak" or rich gravel, is from 2 to 6 feet and is said to carry from $2.50 to $13.00 per cubic yard.

From data furnished by writers, it would appear as if ice rivers were the factors that disintegrated and transported the gold to those places where it is now found in Siberia and the Klondike.

Between Cape Nome and Point Rodney, Alaska, for a distance of 25 miles, there is an ancient sea-beach that extends back from the ocean a distance of 2500 feet. It contains a placer deposit, with the pay streak 20 feet above sea level, and from 12 to 32 feet below the surface. From the westerly base of Cape Nome, gravel terraces or ancient beaches rise gently to the north and west, forming ridges that extend 4 or 5 miles inland, until their highest elevation, 250 feet, is attained. Between the ridges there are broad valleys that contain tundra or arctic swamps.

Within the last two years the tundra have been prospected and found to contain very rich placer ground. The pay streak is situated just above bed rock at depths varying from 60 to 130 feet. The ground is as a rule frozen to bed rock, but here and there soft places are encountered, due to some underground circulation of salt water.

The bed rock is sedimentary, and contains gold, ruby and black sands that have been cemented in mud by quartz solutions.

Above the bed rock is ruby, magnetite and quartz sand in fairly evenly sized grains, with some quartz pebbles and gold in flakes. This is the pay streak, and above it to the surface are alternate layers of rounded beach gravel and gray sands.

The pay streak is so marvelously rich in the tundra mines of Little Creek on the Seward Peninsula, it is difficult to account for its origin. One theory is that glacial rivers formed the placers by pushing ahead of them the gold and gravel they ground from the rocks they passed over. Another theory is that the receding waters of the Arctic Ocean and possibly subsequent upheavals produced the placers and the beaches. The third theory is that the ocean waves pounded the coast rocks to pieces and concentrated the gold.

In the eastern United States along the coast, glacial action so denuded the rocks that no placers are reported north of Maryland. From Maryland to Georgia, to the east of the Appalachian Mountains, placers are found. In several localities in Virginia gold is found in clayey, sandy soil, containing also quartz pebbles. The gold vein is not far from such deposits. In North Carolina, the gold is found in decomposed Cambrian schists; and in chemically altered rocks. In such places the rock contained quartz stringers that carried the gold. In some of the brooks that have worn away these deposits free gold is found, but the placer itself has not been disturbed by either glacial or water erosion.

In the early days it was assumed that all placers if traced up would lead to the discovery of their source or the "mother lode," and that the lode would prove infinitely richer than the placers. The lode may frequently be found, but nine times out of ten it does not contain free gold in anything like the proportion or the size of grains that the placers would indicate. Original rocks are found that will not show gold to the naked eye, and at times not with a magnifying glass, yet the placers from these rocks contain both fine and coarse gold.

The Edith mine in Catawba County, North Carolina, is an instance where gold is found in nuggets through the disintegrated rocks, there being no vein present. The Sawyer mine in Randolph County, North Carolina, and the Morrisville mine in Virginia are instances where the gold is so fine it cannot be seen with the eye, yet both have placers that show coarse as well as fine gold. The Breckenridge, Colorado, placers, have produced considerable wire gold, and the mother lode is traced with reasonable certainty; yet the vein has never paid when worked.

The gold ore that took the prize at the Centennial came from Louisa County, Virginia, yet the placers in that vicinity, or those near the Cabin John mine, Maryland, do not show such large nuggets. From what has been said it is difficult to tell from the size of placer nuggets just what kind of gold will be found in the mother lode. Gold was discovered in California Gulch,

Colorado, as placer, but when traced to Leadville at the head of the Gulch, the greater part of the valuable deposits were silver-lead.

The richer placers of California did not lead to important quartz lodes, and the Comstock lode, although rich in silver and gold, did not show gold in the vein.

The history of placer mining is such that to trace up a placer and find a rich free milling-ledge is not the rule, but generally the exception. Rich veins have been discovered where no traces of placer could be found, except in the grass roots directly over the vein, and again placers have been found where no vein existed. That gold should be found in paying quantities and in sizes from flour to nuggets in placers and not in the mother lode in similar sizes, seems mysterious, and, if some placer miners are to be believed "it grows." This latter statement miners will illustrate as follows: In 1875 California Gulch was washed and $20,000,000 in gold was recovered; when it was abandoned one could not make wages. Of course, some seeds were left, and in 1885 it was washed again and $5,000,000 recovered, the inference being that it grew. The explanation that possibly millions of tons of gold-bearing material were concentrated in that gulch by Nature, and that the second washing was only the leavings of the first, will not satisfy the miner, who will probably paraphrase Job and reply, "There are veins of silver; but the place for gold is where you find it."

Assume that two small pieces of gold are in contact and resting on a rock, and again assume that water moves a fair-sized stone so that it strikes the gold and welds the two pieces. This assumption is not untenable if one will consider how easily the dentist welds gold when filling a tooth, and it may account for "gold growing."

Water has in some instances carried the gold many miles down a sloping hard river bed, and eventually deposited it; in other instances the distance traveled has been short.

Gold may move slowly down the sides of a mountain and be found in streaks parallel to the mountains trend.

Placer deposits are found in narrow streaks or in wide belts, according to their location, the manner in which they were formed, and afterwards acted upon by Nature's forces.

Gulch placers are necessarily narrow, old river beds are much wider, and in some cases where there have been upheavals and severe dynamic disturbances followed by torrents of water they are quite wide. The original placer may have been spread broadcast over an ancient sea or lake bottom that was afterwards raised by an upheaval and formed what is now known as a "dry placer," that is, a placer in a locality where there is no water.

Dry placer areas are quite extensive, and are found in deserts and in some of the arid counties of New Mexico, Arizona, Nevada, Lower California, Mexico, Australia, and possibly India. Gold is not equally distributed through a placer, owing to the current shifting when the placer was formed, for which reason one miner may be working in pay dirt while another a few feet away is not making wages. The miner of experience is very cautious in following up his pay streak, and will dig to the right and left of a rich spot before going ahead.

Gulch mining is particularly uncertain, and frequently the richest dirt is along the sides, and not in the center of the gulch.

No one can with absolute satisfaction explain the various causes for the distribution of gold in placers, for which reason the ground must be systematically sampled.

At times the gold will be uniformly distributed through the dirt, at other times it will be in bunches with barren dirt above and below, and in some cases it will be concentrated in spots in a bench termed a pay streak. It has been found from grass roots to bed rock but in the majority of cases the richest deposits are in the latter situations. If there be depressions in the bed rock where gold can accumulate the pay streak may be very rich at times. Where bars were formed in ancient streams by eddies, the gravel may be very rich, hence the necessity for carefully following the pay streak in bench mining.

The thickness of placers will vary from a few inches to several hundred feet, and where one deposit may have but a single pay streak, another may have several. From what has been said the reader evidently understands that the composition of the placer dirt will vary in different localities. The easiest dirt to wash is sandy gravel. Hardpan, composed of gravel cemented with clay, is more difficult, for if wet it is hard to shovel, and if dry it is difficult to pick. Usually fine sand is not as rich as coarser sand, and to this may be attributed some of the failures in dredging river bars.

The character of the gold found in placers is as different as the placers. Coarse gold in nuggets is the easiest to recover, on account of its weight and shape. Flake gold is not so easy to save as nugget gold; however, its weight will make it sink as soon as it is cleaned sufficiently to prevent the water moving it. Leaf gold resembles flake gold, but is much thinner and lighter and consequently will sink with difficulty, muddy water seemingly having sufficient buoyancy to carry it away. Flour gold is very fine gold, that may make the assay of a deposit run high; at the same time it is difficult to save, and very often it cannot be saved by the hydraulic methods practiced for nugget gold. The black sands that are usually found with placer gold carry considerable gold and frequently platinum; therefore, wherever it is possible they should be saved.

PLACER PROSPECTING.

The object of prospecting placers known to contain gold is to obtain their average value per yard of material, their depth to bed rock, the proper methods to follow in order to recover the gold, and from this data calculate the anticipated profits.

The sampling of placer gravel-beds for hydraulic mining is a matter requiring considerable research before arriving at a conclusion.

Some persons may be satisfied with washings from ten or a dozen small holes, and from these calculate the value of the placer ground; one Klondike estimate was based on a 16-foot hole and a 6-foot drift for a 1000 × 500-foot placer. Such an estimate may suit in the Klondike, but not in this country, and is unreasonable. Mining engineers will not attempt it; promoters will not base calculations upon such exploration, consequently the schemer with a more extensive knowledge of the dictionary than of gravel deposits is the only one who can calculate in this manner.

To determine the value of a gravel bed in a new location it is necessary to examine the topography of the country, and calculate from exploration its length and breadth; the next important matter is to ascertain the depth of the deposit to bed rock; finally, the position of the channel and its course, considering that gold will naturally be deposited along the channel, and will in the lower alluvions conform to bed rock. By this system, the necessary data, as a basis for calculating the amount of gravel, is obtained, but the value of the pay streak is lacking.

If the deposits be deep and bed rock cannot be drifted on, shafts should be sunk and the pay streak worked across the entire bed rock, and for a considerable distance along its channel.

The washings from the dirt so excavated will give the value of the pay streak per cubic yard as closely as may be determined. The thickness of the pay dirt, its length and breadth being now known approximately, its value may be calculated by the known value per cubic yard.

This, however, does not give the value of the whole mass of gravel per cubic yard which is to be worked. The total value being calculated for the pay streak, the total number of cubic yards of the deposit is divided into it, thus giving the unit value per cubic yard of gravel to be worked.

The value of gravel per cubic yard should not be calculated in any other way or expressed in any other terms than in the unit value relative to the whole mass.

In thin deposits tests are made by a series of shafts, or rather holes, across and lengthwise of the deposit so as to cover the entire area. The average returns from such tests will give an approximate value of the placer, and if one test is at fault, it is generally on the safe side. The length, depth and width of the deposit is thus ascertained, and the yardage and average value as determined by the test holes calculated.

In case of more than one bench of pay dirt, the two methods may be necessary to express the unit value of the deposit.

There are two more factors which must be ascertained before the deposit can be valued as a proposition for investment. The first is water in ample supply to break down and sluice the deposit, and the estimated cost of leading the water to the gravel bed. This requires engineering skill, and embraces not only the survey and route for the water course, but reservoirs and dams, and the estimates for their construction, which of course cannot exceed the value of the placers to be worked.

The second factor is that sufficient fall must be had immediately below the workings, to carry away the barren dirt, or at least prevent its accumulating so as to be troublesome.

The basis for the work may be now said to be complete, if the different factors are within reasonable limits of the estimated values; but care should always be used in overestimating work in a rough country, and, in gold-washing, underestimating the gold to be obtained, since matters in the first instance cannot always be definitely calculated beforehand, while it is not generally a rule that the savings are as close in sluicing as in pan or rocker washing, which are generally used to collect the gold when testing deposits.

Until recently it was customary to test placer deposits by sinking shafts at certain marked intervals on the surface. These shafts were put down to bed rock and were about 3 × 3 feet or 4 × 6 feet in area according to depth. It was not possible to make many such shafts on a property, owing to the expense and slowness with which the work progressed, and whenever quicksand was met the hole had to be abandoned after considerable work had been performed.

By the use of the Keystone Drilling Machine prospecting may be accomplished very much easier and quicker than by shaft sinking. The Keystone drill is shown in Fig. 1 to be a portable American drill rig. The drill cuts through the dirt, and is followed closely by a drive pipe. As soon as the earth is broken by the drill, water is added to form muck, and this should be kept liquid. Whenever a certain depth is reached, and the casing pipe is driven down, a vacuum sludge pump is lowered in the hole, to clean out the muck, and this it does effectually. The muck from the sludge pump is dumped into a rocker and washed for gold.

In some cases a record of the drill-hole is kept foot by foot, and the sludge from each section is washed separately, care being taken to save the very fine gold. The records from each hole will show rich and barren streaks, and the depth to bed rock where drilling ceases.

FIG. 1.

While the drill is the more economical method of testing placers it is claimed by some to furnish too high values. Mr. D'Arcy Weatherbe [1] contends "that shaft sinking is the most effective practical system of prospecting placer ground." He further says: "Shaft sinking is feasible in the large majority of cases, is cheaper than drilling,

[1] Mining and Scientific Press, Nov. 3, 1906.

above water level, and in 90 per cent of the cases is not increased in undue proportion to results.

"The cost of shaft sinking by the 'China' method, varies between 90 cents and $1.50 per foot to water level, and the increase below this depth varies with the amount of water." The cost of prospecting with a drill is $2.50 per linear foot. Mr. Downie of the Keystone Driller Company replies to Mr. Weatherbe in an interesting article.[1] After furnishing a history of the drill in prospecting he says:

"It is practically impossible to sink a shaft, by the China method or any other, through boulders and quicksand, and take out an exact area. The inevitable influx of silt and sand from the surrounding embankment may completely vitiate the test. On the contrary, if carefully done the $7\frac{1}{2}$-inch hardened steel drive shoe usually cuts an exact area from top to bottom, and the contents of this test tube may be taken up with exactness and certainty. Of course, some allowance must be made when the ground is entirely composed of large boulders, but even here the chances are in favor of the drill, for it will break a round hole right through them. The presence of water, quicksand, flowing silt, etc., offers no resistance whatsoever, but may and does require a little different method of handling the drill and driving the test tube.

"The question which Mr. Weatherbe raises was brought out and conclusively settled eight or nine years ago at Oroville, by a system of experiments. Drill-holes were first sunk and the values obtained were care-

[1] Mining and Scientific Press, Dec. 8, 1906.

fully computed; then, without removing the test tube, a shaft was sunk around it to bed rock and the values from it likewise accurately computed. The two sets of values were then compared and my informant stated that the value per surface area of the *shaft* was somewhat the larger, but the dredge afterward proved that the value determined by the drill was nearer correct.

"As to the difference in cost of prospecting by these two methods, the question is perhaps an open one. No general conclusion that will absolutely apply to all localities is possible. Shallow tests in dry, clayey ground might in some instances be made by shafts at a lower cost. But in frozen ground, such as is found in Alaska and Siberia, or in submerged quicksand, such as is liable to occur anywhere, the cost by the drill method has been proved by hundreds of experiments to be far lower. On one occasion, in Oregon, with the assistance of two men, in a quicksand swamp where water stood within 12 inches of the surface, the writer sunk holes 88 to 91 feet deep at the rate of one in 16 hours. In average ground it is common to make a 35- to 40-foot test in 10 to 12 hours with a crew of three men.

"But first cost is of little significance. *Accuracy* is the thing that counts. And without going further than the complete and invariable success of the operations at Oroville, the verdict is for the drill method."

The land to be prospected in old river beds in California is divided into blocks of 5 or 10 acres, and a shaft is sunk or a hole drilled in the center of each block. This would not be a feasible method of testing new ground, and one hole at least should be put down to each acre. The ancient river beds should be prospected by driving drifts from the shafts, provided the drifts cannot be driven from the rim rock on the side hill. They may also be tested by drill-holes even if they are from 400 to 500 feet below the surface. No shaft or drill-hole is complete until it reaches bed rock, and care must be used to distinguish false bed rock from bed rock. A record of the ground passed through should be kept and a section made for reference and comparison when drilling or sinking other holes. For this reason the dirt from each advance is washed separately, care being observed to save the fine gold. A cross-sectional map constructed from data in this way will show rich and barren ground, false bed rock, boulder ground, clay, and water level, thus furnishing a guide not only to the value of the ground, but the conditions that will be encountered in working.

The outside diameter of the pipe is taken when calculating the cubic contents of the drill-hole; for example, a 6-inch hole would have a 6.5-inch casing pipe hence $6.5^2 \times .7854 = 33.1813$ and this would give

$$\frac{33.1813 \times 12}{1728} = .23 \text{ cubic foot}$$

in each one-foot length of drill-hole. Owing to some material being sucked by the pump from under the end of the pipe, the values are considered high by this method of calculation, and "Radford's factor" *viz.*, — .27 cubic foot per foot is adopted in Oroville, California, as giving fair results. At the conclusion of drilling the values of the various assays are added and the average of the whole obtained by dividing by the depth of the hole in feet. Unless it is desired to keep each bench separately, the gold from all the pannings is collected by quicksilver. The amalgam formed is treated with nitric acid to remove the quicksilver, after which the gold is annealed, dried, and weighed. From the weight of the gold, the cubic contents of the hole, and the value of the ground per cubic yard in this place is calculated in cents.

Where a number of holes have been drilled, the value of each hole in cents per cubic yard is multiplied by the depth of the hole. The sum of these products is then divided by the sum of the depths of all the holes, to ascertain the average values of the ground prospected. In dredging operations the drill is used advantageously to direct the course of the dredger, drilling for this purpose being carried on continuously. It is claimed that very few placer propositions prove unremunerative where thorough and systematic prospecting has been prosecuted.

CHAPTER II.

HYDRAULIC MINING was once defined as a method of mining in which water broke down gold-bearing earth, transported it to sluices, and separated the gold from the earth.

The definition is not sufficiently broad, as hydraulic mining when applied to gold-bearing earths not only breaks and transports, but washes the material and permits the gold to separate by its greater specific gravity. It is also a concentrating and sluicing process. The process of mining and transporting by water can be applied to coal, iron ore, salt and possibly other minerals, for which reason, the term hydraulic mining should not be confined to gold alone. In view of the scope that the term covers, the suggestive term "hydraulicking" is applied to gold mining, and the term hydraulic mining used to cover all materials mined by the use of water.

Hydraulic mining frequently requires the services of civil, hydraulic, mechanical and mining engineers to install a plant, or at least an engineer who is able to combine those branches of the professions mentioned that enter into the business.

Before any attempt is made at engineering the gold-bearing ground should be prospected carefully, in order to ascertain its extent and value. Several million dollars have been expended in overcoming difficult engineering problems to wash dirt that did not contain sufficient gold to pay the cost of the plant, which fact emphasizes the need of thorough prospecting.

The value of the property having been determined, it may be necessary in order to work it successfully to construct dams for storage reservoirs, and a combination of flumes, ditches, and pipe lines, extending from one to one hundred miles in length, and in addition it may be necessary to tunnel mountains, span chasms, siphon across valleys, place flumes on high trestles or suspension bridges, and possibly bracket them to the sides of high cliffs. From what has been stated the reader will understand that hydraulic mining may in one case be simple while in another it may be intricate and difficult.

It is only possible to describe many matters entering into the subject in a general way, while the most important are described in detail.

The use of water for mining dates back to King Solomon's time. Agricola informs us that fire was used to heat the rocks, and then cold water was thrown on them to spall them.[1]

In quarrying where seams exist in bedded rock, and where explosives would be apt to shatter the rock being quarried, water is employed with wood.

[1] While recently in Cuba, the author heard that Cubans used to mine an ore, burn it and then wash it in pans. Upon investigation a calcite vein containing gold was found, and since the Cubans had no machinery this method answered their purpose. A little of the ore was treated this way and gave good results, showing that the ancients were not so slow where gold was concerned.

The method followed is to drill a series of holes back from but parallel to the face, on the line of cleavage. Into these holes dry wooden wedges are driven. These wedges, on being wet, expand and split the rock as desired. The plug and feather generally in use in such cases does not always answer as well as the wedges mentioned.

The danger which arises from the use of gunpowder in gaseous coal mines has produced two classes of expansive cartridges which depend upon water for their utility. The coal is undercut in the usual manner, and holes drilled in the section to be broken down.

1. Into these drill-holes cartridges of compressed quicklime are inserted, after which they are moistened, then tamped. The water used to moisten the lime causes it to slack, expand, and generate steam; this combination breaks down the coal. The economical value of this novelty has not been fully established in this country. The number of drill-holes and lime cartridges would possibly bring the cost of the process up to that of powder; however, the smaller undercut, and the reduction in the amount of slack coal produced, compared with powder, may counterbalance previous objections. The distinctive advantage which this process possesses is the avoidance of explosion in mines which are subject to outbursts of gas.

2. The water cartridge of the second type is also intended for use in fiery coal mines.

It is a metal wedge, so contrived that upon the application of hydraulic pressure it will expand.

To break down the coal a series of wedges are connected, so that when the pressure is applied it is uniform on all. The cartridges being indestructible may be used over again. They have not come into general use in this country. Cartridges of this description, if they could be used from water pressure at the mouth of some metal mines in the West, would be a great blessing, in preventing the fouling of air and loss of life, not to mention economy in the matter of powder, time, and fuse. Their use would be limited to overstopping.

Salt mining uses water in practical ways as follows:

1. As a solvent. For this purpose a series of bore-holes are drilled from the surface down into the deposit by percussion or diamond drills.

Water is then run into the holes and allowed to become saturated with salt, after which the brine is pumped out and more fresh water added.

By a series of these bore-holes near together an underground water course which connects the holes is soon formed in the salt bed. Nitroglycerine fired in the holes will shatter the rock and is useful in hastening the connection. The water, after circulation is established, flows continuously from the surface into one hole, and is pumped out at the same rate it enters from another hole. The working is now permanent, one bore-hole supplies the water, and another is fitted with a deep well-pump to remove the brine.

This method has advantages, in some instances, over any other method of mining salt where the material is to be broken down, hoisted, dissolved, and then concentrated. It also offers the further advantage of leaving the impurities in the mine, and brings the article sought in the proper concentrated form for refining to the vats.

2. The hydraulic mining termed "spatterwork" originated in the salt mines of Europe, where it has received considerable attention. The water used for mining is given a gravity pressure and ejected from a nozzle having a number of small orifices. The water from this nozzle strikes against the salt deposit and wears it away; at the same time, in flowing away it dissolves the salt, leaving the worthless débris to be broken down or removed. The brine is then collected by gravity in sumps or subterranean reservoirs, from which it is pumped to the surface and evaporated.

Spatterwork can be employed in salt deposits for sinking shafts and winzes from a higher to a lower level, or making "rises" from a lower to a higher level.

FIG. 2.

Gangways or rooms may be driven in salt deposits by the method crudely shown in Fig. 2. For side cutting, the main supply pipe for water has coupled to it, by a hose, a standpipe, *SP*. This pipe is wedged between the roof and floor, in an upright position, with the orifices directed toward the face. The water jets wear away the deposit by solution and abrasion, and the deposit recedes from the orifices of the water jets until the projective force of the water has reached its limit. The water is then turned off and the column pipe placed in another position, where the water by its projective force, together with its solving action, can perform more effective work.

The same illustration shows the method of undercutting the deposit of saliferous clay. The spatter pipe is placed upon the floor and is moved forward to deepen the excavation, or laterally to widen it. The undercut having been made, the clay is easily wedged down

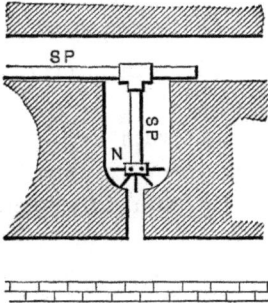

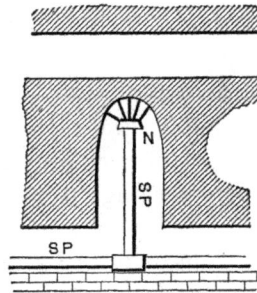

FIG. 3.   FIG. 4.

where it may be acted upon by a stream of water which takes the salt into solution and leaves the barren dirt. The quantity of water is limited to the capacity of the pumps and that necessary for saturation of the brine. Water may, in some instances, be used on one level and be permitted to flow to the next lower level, and so on, thus attaining the requisite saturation before reaching the pumps and sumps.

Wherever the latter conditions prevail, winzes or risers may be made as roughly sketched in Figs. 3 and 4.

To sink the winze, it is necessary to drill a bore-hole from the level above to the level below, to allow the escape of the water discharged from the nozzle N. The water from the supply pipe on the upper level acts by gravity, and propels the water from the jet holes in the . nozzle against the sides of the shaft. It is evident in this instance that the action of the water increases its projective force with depth until it reaches its maximum when the lower level is reached. Fig. 4 shows the method of working out a "rise." To facilitate this latter method, water is brought under pressure greater than the height to be driven, as it decreases in projective force with height.

Mr. Oswald J. Heinrich stated that with a 21-foot head of water, and side cutting from a spatter pipe having twelve brass orifices $\frac{1}{2}$ mm. diameter, the advance was 0.6 square feet per minute, with 1 cubic foot of water per minute. One man attends to 12 spatter pipes in a 12-hour shift. This rate of excavation is in round numbers 5184 cubic feet per day, with 8640 cubic feet of water and one man's labor, thus comparing favorably with any hydraulic mining, as it is .052 cents per cubic yard for labor, and not as high in amount for water as gravel mining generally.

Iron ore deposits of an alluvial character, such as are the "brown ore" deposits of Virginia, can be worked to great advantage by "hydraulicking" if situated on side hills. In such instances the ore is disseminated through clay with barren rocks in such a manner as to need both concentration and washing. It may be necessary to wash ten tons of material to concentrate one ton of ore. The cost of excavating and handling such lean iron oxide deposits would make the bed commercially unprofitable, if freight must be added; it has, however, been practically demonstrated to be more economical to burn fuel and pump water uphill and hydraulic than to work by the former method. To illustrate this more fully: to pick, shovel, and transport the material to the washer, wash it, and load it on cars, will cost, for 10 tons, $2.00 — *i.e.*, one ton of iron ore.

To accomplish the same work with water having a head of 50 feet will cost 75 cents per ton of iron ore. The hydraulic system materially lessens the work to be done by the washer, as the ore becomes freed in a measure from clay as it travels through the sluices to the washer.

There is one more system of water mining made mention of by Pliny in his "Natural History." It has been practiced somewhat in this country, and is termed "booming."

The process of "booming" is to make a dam and collect water; whenever the dam is full the gates are opened quickly, allowing a torrent of water to rush down the hill and upset matters generally. The water, having done its work, is led through sluices which are nearly on a level at the foot of the hill; in these sluices the gold washed out of the soil is collected.

Booming has some advantages which are not to be

overlooked. If there is little water and little working capital the method will be found very serviceable, or if there is considerable water and little working capital it again appeals to the miner. In some cases where there is abundance of capital the method is adopted as the one most feasible for placer mining.

Booming will wash out a large quantity of material, and in its operation is an imitation of a cloud-burst rushing down a ravine. Where there is top dirt above a placer, booming affords a quick and easy method of removing it, provided the dirt is not hardpan and cemented gravel.

Float gold and leaf gold cannot be saved if booming is practiced, and only partially saved by other hydraulic methods.

In Colorado at the Alma and Fairplay placers a system of hydraulic mining termed "ditch waterfall" and "flume waterfall" mining is practiced. At these places there is plenty of water, and this flowing through ditches wears away the earth. The water in the ditch naturally cuts its own channel, thus forming narrow ravines and gashes in the deposit that are useful in assisting the water spurted from nozzles in tearing down the bank. By shifting these ditches or by turning the water into other ditches, considerable space may be covered and the earth washed down to the sluice boxes without any cost of attendance.

This method combined with the pipe work practiced at these places forms the most satisfactory system of hydraulicking. Unfortunately, however, it cannot be followed at every placer mine.

In the early days of anthracite mining, great difficulty was experienced in the preparation of coal for market. All bone coal, or coal frozen to slate or rock, was thrown on the rock pile; and in addition all coal smaller than chestnut size, that passes over a screen with ¾-inch mesh, but through a screen having a mesh 1⅜ inches square, was discarded. This waste accumulated so

FIG. 5.

fast that the culm piles throughout the three anthracite fields became veritable mountains. With the increased demand for coal, the attention of coal operators was given to preparing smaller sizes than chestnut for steam purposes. In 1867 pea coal was first utilized for fuel; in 1878 buckwheat was shipped on a small scale, but as soon as McClave's rocking grate and the Wooton or camelback locomotive were introduced the demand increased rapidly, until at the present time

No. 3 buckwheat or barley size is prepared and shipped in large quantities. When the demand for small sizes became greater than the mines could produce, attention was turned to the utilization of the culm piles. These are mined by water; in fact, hydraulic mining is now carried on to a larger extent for mining coal in north-eastern Pennsylvania than for mining all the other minerals in the United States. The stream of water from a nozzle washes down the coal into a sheet iron trough placed at a slight inclination. The trough connects with the washery where the coal is prepared for market, or with a swinging scraper line such as that shown in Fig. 5 leading to the washery.[1]

Dredging for coal in the Susquehanna River is also carried on from Wilkes Barre to Sunbury, Pennsylvania. The coal found in the river has been transported from the waste dumps at the collieries and from the washeries adjacent in the river.

[1] M. and M., 1903, June. A. I. M. E., Nov., 1905. George Harris.

CHAPTER III.

DEVELOPMENT OF PLACER MINING.

In the early days the ancients depended on placers for their supply of gold. As they had practically no machinery suitable for quartz mining, it may be assumed that the Egyptians previous to Herod's time practiced some form of hydraulic mining. The Romans sluiced for gold, and according to Pliny the shores of Spain were added to by booming. One English writer states that nine-tenths of all the gold has been recovered by hydraulic methods, while an American writer declares that over seventy-five per cent of all the gold mined has been derived from working gravel beds. Probably nine-tenths of all the gold recovered at a profit has been taken from placers. While placers are not as rich ordinarily as veins, and while they cover vastly greater areas than vein formations, nevertheless the gold is more easily recovered from them. This is due to Nature's pulverizing the rocks and concentrating the gold, thereby doing away with underground mining, crushing, milling, or smelting, items which add so materially to the cost of production that vein mining frequently pays only expenses, and more frequently shows a debit balance on the ledger.

The pan, cradle, and sluice were first introduced in the Southern States before gold was discovered in California, but hydraulicking as now practiced was developed in California.

Panning. — The ordinary gold pan of the prospector, while very useful is an imperfect appliance in which to save fine gold. While colors of gold may be detected by the pan, it is very difficult to collect them free from black sand, consequently the pan is useful to placer miners only for nugget gold, unless they use mercury, and this they seldom do. In tracing up gold deposits the pan has no equal, particularly deposits that show free gold. The Spaniards introduced the batea or wooden pan into Mexico, where they are still found to some extent, when sheet iron pans are not available. The most expert panner the writer has ever seen was a Mexican Indian who used a small 10-inch frying-pan with the handle knocked off.

The ordinary sheet iron gold pan, from 16 to 18 inches in diameter, will hold from 15 to 25 pounds of dirt, and with its load will require the use of both hands during washing operations. A smaller pan 10 inches across the top will hold from 3 to 5 pounds of dirt and can be manipulated with comparative ease, and is, therefore, better for prospecting.

A good placer miner, by washing continuously ten hours, can pan from one-half a cubic yard to one cubic yard of dirt, depending of course on the character of the dirt and his nearness to water. If the ground is loose and contains stones of the size of one's fist, more can be washed than when the ground is fine or is cemented material. Ordinary gravel, as found in placers, will probably average 135 pounds per cubic foot; at this figure 27 cubic feet would weigh 3645 pounds. Assuming that each pan washed contained 15 pounds, then it would require 243 pans to wash a cubic yard.

A good days work for a placer miner under medium conditions is 100 pans of dirt in 10 hours. It is difficult to describe the motion given to a gold pan when washing

FIG. 6.

dirt; the object, however, is to separate the gold from the material with which it is associated. The placer dirt is shoveled into the pan until it is heaping full. The pan and its contents are then submerged in water, to loosen the material. The large stones are washed first to remove any adhering dirt that may contain gold, after which they are thrown away. The contents of the pan are then kneaded with both hands, to break up clay, and float the mud away. When there is nothing but sand and gravel in the pan, the panning operation commences and is continued until only the heavier particles remain. If a little clear water is now added the gold in the bottom of the pan will show. In most cases only the gold is saved; however, the black sands may be so valuable it will pay to save them, particularly if there is flour gold.

Gold frozen to quartz is frequently found when panning. This rock should be pulverized and treated as in pan assaying. The pan for this purpose should be black, of Russian sheet-iron, and of the shape shown in Fig. 7.

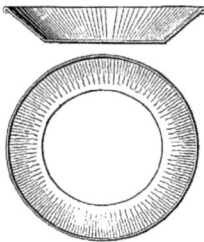

FIG. 7.

The pan is held firmly by one hand, some water is then poured on the pulverized ore; the other hand is used now for shaking the pan in a gentle but rapid manner. The powdered ore being gathered to one side, the heavy grains of gold descend through the sand to the bottom of the pan and settle. After shaking the pan a few minutes, it is to be moved so as to produce a gentle current in casting off the water. This will carry off some of the sand and diminish the quantity in the pan. Fresh water is now added, and another portion of sand washed away, this operation being repeated until nearly all the sand has been washed from the pan. A little water being retained in the pan, the concentrates are moved around by inclining the pan, and giving the water a rocking motion. The gentle current produced by the motion will float the sand away and leave the metal in view. Assaying by the pan is not accurate, as only the coarser particles are retained, the finer going off with the sand. At times it is customary to rock the pan back and forth with the last water slightly and then make a line with the material remaining by inclining the pan to one side. The gold being the heavier, remains at the point of the line in what is termed a pencile. If a batea with a hole in the center has been used for the operation, the gold may be separated from the sand by pushing it through the hole, after it has been collected in the center by a rotary motion.

The Mexican batea (Fig. 8) is a good tool for placer miners, but it does not possess advantages over the iron pan, except, perhaps, in the matter of collecting sulphurets in sample assaying. The wooden bowl

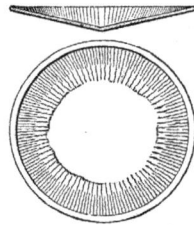

FIG. 8.

is given a steady circular shake without revolving, alternated with a reciprocating motion, which settles the heavier mineral in the center of the bowl; on inclining it the sand flows to one side. In washing they are filled with the dirt the same as pans, immersed in water, and stirred by hand; a circular motion is given to the bowl, which is also slightly inclined, allowing the sand to wash over the sides. The gold sinks to the bottom and clings to the sides of the batea, which requires, generally, more care in manipulation.

To work either the pan or batea requires care and experience; and some become very expert in their use.

The Rocker. — To do away with tedious panning and to increase the quantity of dirt that could be washed in a given time some one invented the rocker.

Rockers are designed in many forms, to suit the ideas of the individual, and often to suit the material to be washed.

The contrivances are rocked back and forth; one swing, however, is longer than the other, the object being to settle the heavy material. In some cases the short swing is brought to an abrupt stop by a block, but more often the man manipulating the rocker decides on the length of swing from the fact that rocking like panning is not an entirely mechanical operation, but requires skill and judgment. The dimensions, like the construction, are varied to suit the ideas of the miner. A fair-sized rocker is about 6 feet long, 24 inches high, and 15 inches wide in the bottom, and 19 inches wide at the top (Fig. 9). The floor of the rocker is given a

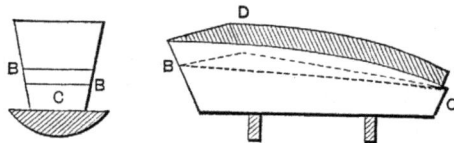

FIG. 9.

slant, with the feed end, *B*, about six inches higher than the discharge end, *O*. This inclination should depend upon the material to be washed and the amount of water available. Fine gold should be given less water and less inclination than coarse gold. Iron bars, parallel to the sides of the trough, are placed on edge, making a grating, known as a "**grizzly.**" These bars have end rests, and if too limber or given to buckling should be stiffened by intermediate rests. The spaces between the bars are from $\frac{3}{8}$ to $\frac{1}{2}$ inch. Perforated or slotted metal plates are more convenient and will answer the purpose as well as bars, besides are more economical if well braced across the rocker. A current of water is let in at the upper end of the rocker, on the ore; this water passes through the grating, carrying the finer material, sand and gold, with it into the box, C. If the gold is fine, quicksilver is placed in small quantities in the box, to form an alloy termed **amalgam.** The light sand in C is swept out by the current of water which passes through the grating at O. At each swing the coarser dirt which does not go through the bars is moved by the jar towards the discharge, O. The jar may not be sufficient to dispose of the coarse material, in which case the miner uses his shovel for that purpose. While rocking is quite effective for coarse gold, there is much fine float gold lost even when quicksilver is employed. This is especially the case when much clay is present as that encases both coarse gold and fine, and since the specific gravity of the two combined is less than for gold alone, the density of muddy water may be sufficient to buoy the fine particles, which float away in the agitated current of water. Mercury cannot reach fine gold smeared with clay, and it may be worth while, therefore, to go slower and use more water to wash off the clay.

Where there is much clay a good plan is to feed the material and water into a trough, and allow the dirt to be moved by the water along the trough and discharged into the rocker. The clay will be washed more thoroughly from the gold by this means, and the latter be given a better opportunity to form amalgam.

Another form of rocker is shown in Fig. 10. This is a box with sloping sides, about 36 to 42 inches long and 16 inches wide, with a rocker near the middle and one near the back. There is a hopper, H, 20 inches square,

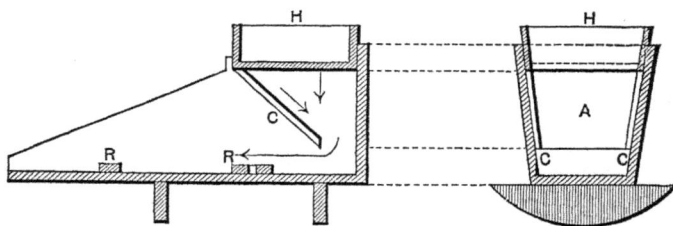

FIG. 10.

4 inches deep, whose iron bottom is perforated with $\frac{1}{2}$-inch-diameter holes. This hopper is removable. Under this hopper, on a light inclined frame, C, a canvas apron, A, is stretched, to form a riffle. The water is poured on the dirt, which is shoveled into the hopper, washes the gold and sand through the screen, after which the coarse material in the hopper is thrown aside and new dirt substituted. The rocker has pieces of plank, R, nailed transversely across the bottom, to catch the gold as the current transports the sand to the discharge end.

The rocker shown in Fig. 11 is used in the South, and those natives who have the gold fever consider it the best apparatus for washing dirt in existence. Although it is a crude affair it is nevertheless effective in the hands of one accustomed to its manipulation. The Southern placer gold is fine, most of it being mere colors, so that pieces from the size of mustard seed up are called nuggets.

Men cannot pan sufficient quantities of this dirt to make wages, but with a rocker can treat from 2 to $2\frac{1}{2}$ cubic yards daily.

The rocker has two longitudinal riffles, a, placed about as in the cross-section. The riffles are about the thickness and width of bed slats, the object being to

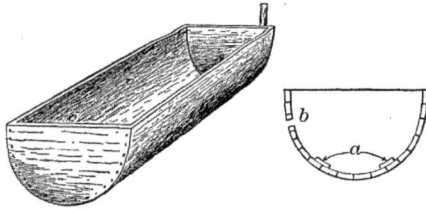

Fig. 11.   North Carolina Trough Washer.

retain the black sands and gold between the two. The dirt to be washed is shoveled into the tub until the bottom is covered. Water is next poured in and the apparatus rocked to clean the mud from the larger pieces of stone. The cradle is then tilted until the dirty water will run out of plug holes, b, after which the larger pieces of rock are raked out over the side. This operation is repeated until the water poured in and agitated remains comparatively clear, and there are but few small stones in the tub. Expert work now begins, the object being to wash the light sands over one riffle and leave the gold and heavy black sands between the two riffles. To accomplish this a quick jerk is given the rocker one way, and then as the water moves to one side it is allowed to come to rest slowly and flow back. The motion is similar to that given a pan when a pencil of black sand is being formed, except that the height of the wave movement is decreased gradually on one side of the tub and increased on the other until all the light sands are on the long-wave side. When this is accomplished the heavy sands are between the riffles, and the gold is picked out. Expert manipulators of these cradles claim that they can save 90 per cent of the gold. They do not use mercury either in the cradle, or in the clean-up, from the fact that it costs money, becomes foul quickly, needs retorting, and must be cleaned before it will amalgamate properly, all of which means extra labor and expense, which they cannot stand, with such small operations.

Combination rockers, such as were used at Gold Hill, North Carolina, are made by connecting several single rockers by rods, the pulp being conveyed to them by a trough. The riffles in these rockers are crosswise of the trough and only one end is closed, making it virtually a rocking sluice box. A woman often furnished the motive power, shifting her weight alternately from one side of one rocker to the other. In many instances these rockers were used as concentrators in conjunction with Chilean mills.

The Long Tom is a short sluice box that is used in place of the rocker in suitable localities. It requires one man to feed it and another to keep it in working order. It is capable of washing 6 yards of ordinary dirt, and from 3 to 4 yards of cemented dirt in 10 hours. The material to be washed is shoveled into the sluice box, H, Fig. 12, and that being supplied with an abundance of flowing water, carries the dirt to the tom. The feed end of the tom is about 18 inches wide, while the discharge end is about 32 inches wide, and terminates in a perforated sheet-iron plate, *P*. As the material enters at *H*, it spreads out until it meets the plate, *P*, where it is immediately riddled and so assorted, that all stuff finer than one-half inch in diameter falls with the water into a second trough, *T*, one end of which is underneath

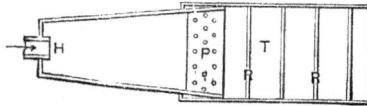

FIG. 12.

the plate. The coarse material is shoveled off the plate, and the lumps of clay and dirt thrown up towards the head so that the water will have another chance to disintegrate them.

In order to facilitate the movement of material in the troughs the latter are placed on timbers or stones to give them a slope towards the discharge. The lower box is furnished with transverse riffles, *R*, which collect the gold that moves with the stream of water. The constant movement of the water beneath the plate keeps the sand suspended and allows the gold to separate and sink by gravity to the floor of the trough. The inclination of the riffle box should be such that the bottom of the trough is covered with a thin coating of mud, and is not scoured by swiftly running water and sand.

Mercury can be used in the riffles to assist in retaining the gold, and the riffle box can be supplemented with another box containing blankets or hides with the hair turned up stream to catch the fine gold. In the latter case a fine screen with not larger than 11 inch mesh holes should be placed above the blanket, in order to prevent coarse sand and gravel from traveling over and wearing it. On Snake River, Idaho, very fine float gold has been saved by such means.

Sluicing is a term used to indicate the process of washing dirt through a channel by means of water. The channel through which the dirt is transported may be a wooden trough, or a ditch cut in bed rock. The sluice is the most important part of any hydraulic mining system, and if it is not constructed properly there will be a loss of gold. The earth may be fed to the sluice by hand, or indirectly by machinery, or it may be washed from a bank by a stream of water that is afterwards led to the sluice.

Fig. 13 shows a hand sluice into which the miners shovel the gold-bearing dirt, previously picking out the large stones and throwing them one side. Similar sluice boxes, although much larger and stronger, are used where mechanical apparatus is employed for excavating.

The water in flowing through the sluice transports the material to the dumping ground, and at the same time washes the gold free from rocks to which it is adhering, and permits it to fall to the bottom of the sluice, where it is caught in artifical traps called **riffles.** There should be arrangements to prevent large stones entering a small sluice as they wear the boxes, require more water for their transportation, or else a heavier grade, than do smaller stones, and are otherwise objectionable. In

FIG. 13

many cases all material to be washed passes through grizzlies that prevent stones larger than 3 inches in diameter from entering the sluice.

In ground sluicing this is not always practicable, as the material is delivered so fast screens would become clogged, for which reason a man stands at the head of a sluice and pulls out the largest stones, and prevents the others from clogging the entrance. Ground sluices are bed-rock sluices, and must be constructed with great care, as much depends upon them. They are objectionable because they can be cleaned up only once in a season, and because all boulders have to be removed out of the way by derricks. The motive power for the derricks is obtained usually from water, and a Pelton water wheel.

Upon the construction and operation of a sluice much of the success of placer mining depends. Steadiness of flow, that is, the quantity of water passing and its velocity should be uniform to secure a maximum settling of the gold. To be sure it is not always possible to prevent crowding a sluice, particularly when caving a bank, but there is no economy in doing so, and generally an experienced pipeman can avoid it. If the bank runs in too freely so as to send large quantities of dirt to the sluice, it may be economical to construct other sluices. The bulk of the gravel and boulders travel down the middle of a sluice, hence its grade and depth are important, yet the fact that a swift current while able to transport heavy material will not permit fine gold to settle must not be overlooked.

The sectional area of a sluice will depend upon the

quantities of water and dirt it is to carry. An ordinary sluice will probably be made of 2-inch planks, 12 inches wide and 12 feet long. This sized lumber will furnish a box 12 inches wide by 10 inches high. The boxes are generally made in 12-foot lengths, and as many placed end to end as the nature of the ground demands. They should be caulked with cotton wick if oakum is not readily attainable, to prevent leakage, and if they are to carry fair-sized stones should have false bottoms, which may be made to act as riffle bars.

The sluice boxes should be placed in a straight line, but if curving is necessary, the outer edge of the curve should have an elevation, to prevent the material from piling up and clogging the box when the direction of the flow is changed. There should be at least one inch elevation for each degree of curvature, but even this will not in all instances prevent retardation after the curves have been passed, making it necessary to give a slightly greater fall below the curve in order to obtain uniform flow of material and clear the curves.

The grade necessary to give a sluice will depend upon the character of the alluvions; large, heavy stuff will require a steeper incline than lighter material. The amount of water at command will influence, in a measure, the gradient, and the sectional area of the sluice must also depend upon it. Heavy material must be covered by water, and a steep enough grade given to have gravity give velocity to the water and exert some little action upon the material; naturally, then, were the sluice broad, 300 cubic feet of water per minute might be required, where with but half that supply of water the sluice must be narrowed or otherwise a very steep gradient given it. Narrowing the sluice would be the most satisfactory arrangement.

The length of the sluice depends upon dumping-ground and its distance from the workings; yet, were the dump close at hand the sluice must have sufficient length to thoroughly wash the alluvions, and break up the cemented gravel, and the clay.

The size of a sluice is to be determined by the amount of gradient at command, the character of the material, and the quantity of water which may be used.

The grade of a sluice will depend upon the fall of the ground to the dump, the character of the material transported, and the amount of water at command. The grade will vary from 2 to 15 per cent,[1] or from 2 to 15 feet in every 100 feet of length, or from 2.8 inches to 21.6 inches per box 12 feet long. The grade should be determined previously by experiment before permanently placing the sluice in position, otherwise there may be considerable loss of both gold and amalgam, to remedy which may require the raising of the whole sluice line, or, if the fall is not sufficient, its lowering. It is important that the sluice has sufficient fall, and a proper dumping ground, also it should be near the level of the ground where the dirt is to enter it.

As low as 1½ per cent grade has been used.

The sluice has advantages over any other system both for collecting free gold and the removal of barren dirt in an economical manner, consequently the attention

[1] Bowie, Alex. J., p. 219.

given to its construction and the work it performs will prove remunerative.

Where gravel is to be sluiced for some distance, the boxes should be stepped, in order to effect a drop that will shake up the material. Sand is apt to sink and move along the sluice in a sluggish and compact manner, and in order to shake it up and permit the heavier particles, *i.e.*, gold and black sands, to move along the bottom of the sluice to a riffle, a step here and there is necessary.

The construction of a sluice box depends for details upon the size required; one 6 × 3 feet would require heavier sills and flooring than one 3 × 1.5 feet.

The sills should be three feet apart and be twice as long as the width of the sluice, provided there is nothing to prevent this construction.

The posts are regulated in height to accommodate the water and material; in this connection it may be stated that wide sluice boxes lessen the water pressure on the material transported, and are, therefore, more satisfactory. However, the kind of material must determine, in a measure, this point. The bottom planks should be made of clear lumber and grooved to admit of a dry pine or other tongue being inserted into the groove. These planks are placed lengthwise of the sluice and securely fastened to the sills. They should be purchased in widths to conform to the total width of the sluice, to avoid expense in transportation and unnecessary delay in placing them. If a tight floor is to be had, half-seasoned plank, not less than 1½ inches thick should be used. The side planks should be worked in a similar manner to the bottom planks, and should extend to the sills, however in some cases one inch boards with battens are used, and in others planks not tongued and grooved but battened and caulked. The side linings should be rather thicker than the side planks, and may be rough plank. Where riffles are inserted they do not reach to the bottom plank; in all other

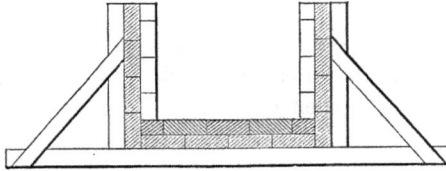

FIG. 14.

instances they should, to avoid wear on the side planks. The posts are braced every alternate sill by means of $1\frac{1}{2}$-inch plank strip, as shown in Fig. 14.

To avoid wear upon the bottom plank, rough plank must be nailed over them. This should be hard wood, if possible; beech or maple will be found to wear smooth and uniform, where oak splinters.

The cost of the sluice box will depend upon the locality, the price of lumber, nails, and labor per diem in that locality, as well as transportation.

The great advantage possessed by sluicing in saving gold is due to the thorough washing the material obtains, but the necessity for the erection of retaining dams to catch the tailings has in recent years greatly retarded the system in California and consequently the output of that state.

Sluices should be at least 240 feet long and set in as straight a line as possible, otherwise fine gold and mercury may pass through. A sluice 500 feet long had frames 5 × 7 inches, with a box 60 inches wide by 30 inches deep inside, it was lined on the bottom with 2-inch planks and on the sides with 1-inch boards. The floor lining was 6 × 6 inches blocks 10 inches long of sawed lumber. Another sluice 2000 feet long was paved with blocks.

Figure 15 shows two sluice-box lines at Bullion, British Columbia, where hydraulicking is practiced.

Transporting Power of Water.—Water will carry dirt in suspension, and thereby increase its density; it must however have a current, otherwise it will precipitate the dirt. If the current is swift and the water deep only the heavy particles will be precipitated, but if the water is suddenly spread out and the current reduced the light particles will fall to the bottom. It is upon this principle that the sluice and undercurrent are constructed; the first being intended to transport all dirt and water that is not too heavy, and the second being for the purpose of recovering light particles of gold that are not sufficiently heavy to sink in a swift running stream. The dirt when in suspension adds to the density of the liquid and hence to its transporting power. This may be better understood by considering the momentum exerted by water moving at a given velocity in feet per second. One cubic foot of water will weigh 62.5 pounds, and if it move at the rate of 10 feet per second, will have a momentum of 625 pounds. The weight of a cubic foot of wet sand is twice that of water, 125.0 pounds. Supposing 1 cubic foot of

FIG. 15.

material and water passing along the sluice to be composed of two-thirds water and one-third sand, the weight would be 82.6 pounds, and the momentum at the above velocity 826 pounds, thus increasing the transporting capacity of the water one-third. The density of the water having been increased one-third, its ability to float material has been increased one-third; or, expressed in momentum (as far as the rock in the sluice is concerned, whose specific gravity relative to the fluid is decreased one-third as compared with water), 1101 pounds. The transporting capacity of such a combination is therefore nearly double that of water alone, hence the coarse and heavy material moves along, not on the bottom of the sluice, but above the bottom and below the water. This combination will move rocks that aid by their movement to disintegrate and wash out the gold from the dirt that may hold it encased or in suspension, and also prevent sand from packing. Heavy rocks will not have the same velocity as lighter, but their colliding has a grinding effect upon the material containing the gold. There are no experiments of such a nature as to formulate a rule by which the gradients being known the transporting capacity can be determined. According to Le Conte, "If the surface of running water be constant, the force of running water varies as the square of its velocity, and the transporting power of a current as the sixth power of the velocity." [1] Friction increases as the square of the velocity and as the cube of the density; however the same liquid will vary in density in the same sluice, to a wide degree. Le Conte says: "The transporting power

[1] Elements of Geology, pp. 19, 20.

of water will be between the square and sixth power of its velocity." According to Smeaton, a velocity of 8 miles an hour will not derange quarry rubble stones, deposited around piers, provided they do not exceed half a cubic foot, except by washing the soil from under them. The transporting capacity of water in sluice boxes will be greater than in rivers from the fact that they are smooth and straight and of uniform width.

The Size of Sluice Boxes. — Sluice boxes should be calculated for carrying capacity, but in order to do this they must be calculated from the quantity of water that is to flow through them. The maximum quantity of water that can be used to advantage in a single sluice is stated to be 1000 miners' inches or 90,000 cubic feet per hour, a miners' inch being legally in California 1.5 cubic feet per minute. If there is more than this quantity of water at command, two sluices should be used, for a greater flow takes the workmen off their feet.

It is a difficult matter to calculate the size of sluice boxes owing to the uncertainty of the size of the material they are to carry, consequently assumptions must be made and calculations made to agree with them.

The area of a sluice is to be calculated in square feet, thus a sluice 2 × 3 feet is 6 square feet in area, and if such a sluice were 10 feet long it would contain, when full of water, 60 cubic feet. If this box were given a slant of 1 foot, it would have a grade of 1 foot in 10 feet or 10 per cent, and the water would flow with a velocity found by the formula

$$v = \sqrt{2\,gh},$$

in which v = velocity, $2\,g$ = 64.32 the acceleration of gravity at the end of the first second; and h the height of fall or grade which in this case is 1 foot. Substituting these values

$$v = \sqrt{64.32 \times 1} = 8.02,$$

or the velocity in feet per second which the water would have at such a grade.

The flow would be, if there were no other interfering factors, 8.02 $\times$ 6 = 48.12 cubic feet per second. But there are other interfering factors such as frictional resistance, which increases as the rubbing surface increases and as the velocity of the water increases.

The rubbing surface is the surface wet by the flowing water, and in the example cited it is 2 + 3 + 2 = 7 feet. This wetted surface or rubbing surface is termed the *wet perimeter* and must enter into the calculations as it retards the flow of water. If the channel has smooth sides such as wooden sluice boxes usually do have, the friction is less than in ditches, consequently the size and grade may be less for the same volume of water.

Before entering into the calculation of the carrying capacity of sluice boxes there are some terms that should be explained.

The area of a sluice box is the number of square feet or square inches in its vertical cross-section. The *area* is found by multiplying the width by the depth; for example, Fig. 16, 1.5' $\times$ 3' = 4.5 square feet is the area of the box; or 18" $\times$ 36" = 648 square inches is the area of the box in square inches. In calculations of this kind inches must be

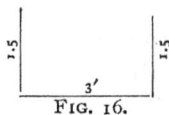

Fig. 16.

multiplied by inches and feet by feet, but either may
be reduced readily to the other; thus

$$4.5' \times 144'' = 648 \text{ square inches;}$$

and
$$\frac{648''}{144''} = 4.5 \text{ square feet.}$$

Another term of frequent occurrence in hydraulics
is *wetted perimeter*. The perimeter of a figure is its
boundary lines, and is measured by the length of those
lines, thus in Fig. 16 the perimeter is

$$1.5' + 3' + 1.5' = 6 \text{ feet.}$$

The wet perimeter of a sluice box is the border of the
box in actual contact with the water, thus in Fig. 16 if
the water was running 12 inches deep in the box, only
1 foot on each side of the box would be wet, consequently
the wet perimeter would be $1 + 3 + 1 = 5$ feet or
60 inches. The water area in this case would be
$3 \times 1 = 3$ square feet.

The **mean hydraulic radius** is equal to the water
area of the sluice, divided by the wet perimeter, for
example, the water area in the illustration is assumed
as 3 square feet and the wet perimeter as 5 feet, hence
the mean hydraulic radius is $\frac{3}{5} = .6$ feet.

The mean hydraulic radius is called at times the
hydraulic mean depth, and **hydraulic radius.**

If a sluice box 12 feet long had a water area of 3
square feet it could contain when horizontal $3 \times 12 =$
36 cubic feet of water. If the box were tipped so that
one end was 6 inches higher than the other, the water
would flow to the lower end with a velocity equivalent
to so many feet per second. If there was nothing to

prevent the flow of this water it would have a velocity according to the formula

$$v = \sqrt{2\,gh},$$

in which v stands for velocity g for the acceleration of gravity $= 32.16$; and h for head or height of fall. Substituting the values of the example in this formula,

$v = \sqrt{2 \times 32.16 \times .5} = \sqrt{32.16} = 5.76$ feet per second as the velocity. To find the quantity of water that would flow with such a head, under such theoretical conditions, all that is necessary is to multiply the area by the velocity or $3 \times 5.76 = 17.78$ cubic feet per second. There are several matters which enter into such calculations, that prevent the theoretical quantity of water from being realized in practice. There must be an allowance made for retardation due to friction; moreover the wetted or water area only has been considered, and when transporting material the wet perimeter will be increased from 20 to 35 per cent of the depth of the water alone.

In rectangular sluice boxes, the wet perimeter is smallest and, therefore, the friction least, when the width of the box is from $1\frac{3}{4}$ to $2\frac{1}{4}$ times the vertical depth. In all cases the boxes should not run more than $\frac{3}{4}$ full.

When the area needed to convey the quantity of water at the given velocity is ascertained, it is necessary to find what form the area shall take.

The formula for finding the area is

$$a = \frac{q}{v},$$

in which a = area in square feet or square inches,
  q = quantity of water flowing in cubic feet per second.

In the former example, the area was assumed and the velocity and quantity calculated to the conditions. This was not the best form of a sluice as the wet perimeter and consequently friction was increased: assume, however, that the area was the same (3 square feet), but the bottom width was $1\frac{3}{4}$ times the height, then to find the proper cross-sectional area for this width;

" *Multiply the given area in square feet or square inches by 4, and divide by 7; the square root of the quotient will be the depth in feet or inches.*"

Example. — What will be the depth in feet of a sluice box having an area of 3 square feet, when the width of the bottom is $1\frac{3}{4}$ times its height?

Solution. — $\dfrac{3 \times 4}{7} = \sqrt{1.714} = 1.309$.   Ans.

If the width is to be $2\frac{1}{4}$ times the depth or side, " *multiply the given area in feet or inches by 4, and divide by 9; the square root of this quotient will be the depth in feet or inches.*"

Take the area 3 square feet as before.

$$\dfrac{3 \times 4}{9} = \sqrt{1.333} = 1.15.   \text{Ans.}$$

This gives a larger area than where the rubbing surface is not considered. Probably the best width relative to the depth is 2 to 1.

Grade. — To create a uniform flow the sluice must be given a grade that will produce the required velocity.

This must necessarily vary according to the material to be transported, since coarse material will require a steep grade or a large quantity of water, while fine material will not require a very steep grade or much water.

The grade given a sluice box varies from 2 to 16 inches, a box being 12 feet in length. The average grade is probably 6 inches for a 12-foot box, where the gravel comes from a bank that is hydraulicked. The grade once determined should be adhered to and only broken at undercurrents.

To calculate a sluice gradient to produce a given flow of water use the following deduced formula

$$v = c \sqrt{2 \ grs.}$$

In this formula

s = sine of inclination or $\dfrac{\text{total fall in feet or inches}}{\text{total length in feet or inches}}$.

c = a coefficient determined experimentally for rough planks to be .8.

g = 32.16, and is the acceleration in 1 second due to gravity.

v = velocity in feet per second.

r = hydraulic mean radius or $\dfrac{\text{sectional area}}{\text{wet perimeter}} = \dfrac{a}{p}$.

Example. — The water area of a sluice box is 2 square feet; the perimeter of the box is 4 feet; the grade is 6 inches in 12 feet; What will be the velocity of flow? What will be the quantity of water passing?

Solution. — $r = \dfrac{2}{4} = .5.$ $s = \dfrac{.5}{12} = .0416$

then by substituting in formula

$$v = c\sqrt{2\ grs} = .8\sqrt{32.16 \times 2 \times .5 \times .0416}$$
$$= .8\sqrt{1.337856} = .8 \times 1.156 = .92 \text{ feet per second.}$$
$$q = va = .92 \times 2 = 1.84 \text{ cubic feet per second. Ans.}$$

In the above example the grade s was given, however, to find a grade that will furnish a given velocity, the above formula is factored until it assumes the form

$$s = \frac{v^2}{c^2\ 2\ gr}$$

Example. — What grade must be given a box whose water area is 2 square feet, and whose perimeter is 4 feet, in order to produce a flow of 1.84 cubic feet per second.

Solution. — $v = \dfrac{q}{a} = \dfrac{1.84}{2} = .92$ feet per second.

$$s = \frac{v^2}{c^2 2\ gr} = \frac{.92^2}{.8^2 \times 2 \times 32.16 \times .5} = \frac{.8464}{20.58} = .0416$$

sine of inclination, and hence

$$.0416 \times 12' = .499 \text{ or } .5 \text{ feet in } 12 \text{ feet. Ans.}$$

In order to find the dimensions of a sluice box that will carry a given quantity of water, the empirical formula

$$v = \frac{\sqrt{100,000\ r^2 s}}{6.6\ r + .46}$$

is adopted by many engineers.

In this formula r is the hydraulic radius, and s is the slope of the sluice, or $\dfrac{h}{l}$.

Example (1). — It is required to compute the dimensions of a sluice to convey 2.82 cubic feet of water per second with a grade of 6 inches in 12 feet, the width of the sluice to be twice the depth of the water flowing through it.

Solution. — Let x = depth of water in the flume; then the width will be $2x$ and the wet perimeter = $4x$.

The water area will be $2x^2$, and the hydraulic radius $r = \dfrac{2x^2}{4x} = \dfrac{x}{2}$, consequently $r^2 = \dfrac{x^2}{4}$. The slope is $\dfrac{6}{12 \times 12} = \dfrac{1}{24}$. As the discharge is to be 2.82 cubic feet per second, the mean velocity will be $\dfrac{2.82}{2x^2} = \dfrac{1.41}{x^2}$.

Substituting these values in the formula

$$\frac{1.41}{x^2} = \sqrt{\frac{100,000 \times \dfrac{x^2}{4} \times \dfrac{1}{24}}{6.6 \times \dfrac{x}{2} + .46}}.$$

Squaring

$$\frac{1.9881}{x^4} = \frac{100,000 \times \dfrac{x^2}{4} \times \dfrac{1}{24}}{6.6 \times \dfrac{x}{2} + .46} = \frac{\dfrac{100,000 x^2}{96}}{3.3x + .46}$$

then

$$6.56073\, x + .914526 = \frac{3125\, x^6}{3}.$$

Multiplying by 3

$$19.68219\, x + 2.743578 = 3125\, x^6.$$

Dividing by 3125

$$x^6 - .006\, x = .0008779.$$

If $\qquad\qquad x = .383$

then $.003156 - .002298 = .0008584,$
which is close enough to the second term of the equation
and hence

$$x = .383 \times 12 = 4.596 \text{ inches.}$$

$2x$ or width of sluice $= 2 \times 4.596 = 9.192$ inches.

$4x$ or wet perimeter $= 4 \times 4.596 = 18.384$ inches or
$1\frac{1}{2}$ feet.

To prove this

$$\frac{4.596 + 4.596 + 9.192}{12} = 1.532 \text{ feet}$$

for wet perimeter.   Area of water section

$$\frac{9.192 \times 4.596}{144} = .293 \text{ square feet.}$$

Hydraulic radius $= \dfrac{.293}{1.53} = .191$ feet.

Substituting these values in the equation

$$v = \sqrt{\frac{100,000 \times .191^2 \times \frac{1}{24}}{6.6 \times .191 + 46}} = 9.399 \text{ feet per second and}$$

since $q = v \times a$ the discharge would be

$$9.399 \times .2933 = 2.757 \text{ cubic feet per second,}$$

which is approximately 2.82 cubic feet per second. All
such formulas are approximate and tedious to work
as trial depths must be taken. Thirty-five per cent
must be added to the width of the sluice to accommodate
the water and material, and 35 per cent should be added
to the depth, because a sluice should not run more than
$\frac{3}{4}$ full, consequently the size obtained by the above
formula should be increased 70 per cent.

Example (2). — It is required to compute the dimensions of a sluice to convey 28.2 cubic feet of water

per second with a grade of 6 inches in 12 feet, the width of the sluice to be twice the depth of the water flowing through it.

Solution. — Let

x = depth of water in the sluice.

$2 x$ = the width of the sluice.

$4 x$ = the wet perimeter.

$2 x^2$ = the area of the water cross-section.

$$\frac{2 x^2}{4 x} = \frac{x}{2} = \text{the hydraulic radius.}$$

$$\frac{x^2}{4} = \text{the hydraulic radius squared.}$$

$$\frac{6}{12 \times 12} = \frac{1}{24} = \text{the slope or grade.}$$

$$\frac{28.2}{2 x^2} = \frac{14.1}{x^2} = \text{the mean velocity.}$$

Substituting in formula

$$v = \sqrt{\frac{100,000\, r^2 s}{6.6r + .46}}$$

$$\frac{14.1}{x^2} = \sqrt{\frac{100,000 \times \dfrac{x^2}{4} \times \dfrac{1}{24}}{6.6 \times \dfrac{x^2}{2} + .46}} = \sqrt{\frac{\dfrac{100,000\, x^2}{96}}{3.3 x + .46}},$$

or

$$\frac{14.1}{x^2} = \sqrt{\frac{3125\, x^2}{9.9 x + 1.92}}.$$

Squaring

$$\frac{198.81}{x^4} = \frac{3125\, x^2}{9.9 x + 1.92}.$$

Clearing fractions, $1968.219\, x + 381.7152 = 3125\, x^6$.

Dividing by 3125 and transposing,

$$x^6 - .62983\, x = .1221488.$$

Assuming the depth of the water 1 foot for x and substituting this value in the left hand number of the equation,

$$1^6 - .62983 \times 1 = 1 - .62983 = .37017,$$

which is greater than the second number of the equation and shows the assumed value is too great. Trying a value $x = .9$,

$$.9^6 - .62983 \times .9 = .5314 - .5668 = -.0354,$$

which is less than the second number of the equation, hence .9 is too small for x and the value must be between 1 and .9.

After repeated trials x is found to be equal to .94635, therefore,

$$.94635^6 - .62983 \times .94635$$
$$= .71828333 - .59603962 = .12224371,$$

which satisfies the required condition, nearly. Hence x or depth of sluice is .94635 feet or 11.3562 inches.

$$2\, x = 1.8927 \text{ feet or 1 foot, 10.7 inches.}$$

To verify the foregoing dimensions:

The wet perimeter = $.94635 \times 2 + 1.8927 = 3.7854$ feet.

Water area $= 1.8927 \times .94635 = 1.79115$ sq. ft.

Hydraulic radius $= \dfrac{1.791156645}{3.7854} = .47317.$

Substituting in the formula,

$$v = \sqrt{\frac{100,000 \times .47317^2 \times \frac{1}{24}}{6.6 \times .47317 + .46}} = 16.136 \text{ feet per second}$$

and $16.136 \times 1.79 = 28.88$ cubic feet per second, which satisfies the condition of the problem approximately. Thirty-five per cent must be added to the depth and width of the box to accommodate the material and 35 per cent to the depth, because the sluice should not run more than three quarters full.

With these calculations made as follows the sluice would have

$$1.8927 \times 1.35 = 2.565 \text{ feet as the width; and}$$
$$.94635 \times 1.35 = 1.28 \text{ feet as the depth.}$$

At large operations there should be at least two sluice lines, in order that work need not be stopped. It is not absolutely necessary to have two sluice lines, but it will permit mining to be carried on when one sluice is out of commission, either for repairs or for cleaning up. Where there is plenty of water and much ground to be washed, the construction of two or three sluices may be advisable.

CHAPTER IV.

Riffles were mentioned under rockers and are merely traps intended to stop gold from moving along the sluice bottom. For coarse gold they are very effective, but for gold containing impurities, gold attached to rock, leaf gold, flour gold, or gold attached to black sands they are not satisfactory savers, unless aided by undercurrents and mercury.

There are many kinds of riffles, some of which are patented, consequently it is necessary to use judgment in their selection, as there would not be this number if it were possible to save all the gold in places. A placer miner after long experience in a certain mine stated that he did not know how much gold was in his dirt, but he knew it was worth 60 cents per yard gross to him, for that was his average saving. Some placers that have assayed rich in gold have proved flat failures as business propositions owing to the physical condition of the gold, and the writer has had several experiences where the value of a sluicing proposition has been based upon the fire assay of the concentrates that included black sands. Sluicing is a purely mechanical operation, and only free gold enters into a proposition of this kind, therefore one must know within reasonable limits what the physical condition of the gold is before attempting work. Coarse or nugget gold will travel but a short distance even if there are no riffles, provided there is an uneven place for it to lodge, however a stone moving along will dislodge it and then it moves to the next uneven place. To prevent this riffles are needed as otherwise the gold will eventually reach the end of the sluice.

Flour gold is very fine and will be held in suspension in muddy water; leaf gold will float along the bottom of a sluice unless it can be stopped in some way, and gold attached to black sand or rock will be lost.

It is no wonder then that there should be a variety of riffles, and all based upon their ability to save gold. Some prefer transverse riffles that have proved effectual in saving gold in operations with which they were once connected; others prefer longitudinal riffles for the same reason, however this rule of thumb will not suffice and as previously stated the operators must use judgment. The writer has saved gold in a longitudinal riffle that seemingly would not remain in a transverse riffle, however he found that by a combination of both he could save more than with either separately.

In some cases, mercury will not save gold, in other cases it will do so, therefore mercury should be used in conjunction with riffles when the latter prove ineffectual in saving gold that mercury will attack.

Coarse gold does not need mercury, while fine clean gold can only be saved by its use, when however mercury is discarded in undercurrents there is a mistake made for it has been proved in rich placers like those of Little Creek on the Seward Peninsula that mercury saved fine gold, that it is customary to run into the tailings. When gold is encased in the oxides or sulphides of other metals, the combination prevents amalgamation, besides lessens the specific gravity of gold to such an extent it floats away in the current. This kind of gold material can not be saved except by concentrating the sands, or griding them so as to expose the gold, and then giving them a lixiviation treatment.

Pole Riffles made of saplings split and nailed to the sluice-box floor are the crudest of riffles, and yet they are useful when better riffles can not readily be obtained. The saplings are placed transversely or longitudinally to suit the ideas of the operator. Where there is much clay in the ground nails are driven in the poles so that they will project a half inch. This breaks up the clay in a measure, and thus frees the gold.

Board Riffles are shown in Fig. 17 to be longitudinal strips nailed to scantlings, that are as long as the sluice is wide inside. This riffle forms a false floor to the

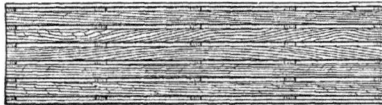

FIG. 17.

sluice and prevents that wearing, at the same time the gold that passes in between two boards is held by the scantlings. Such riffles are readily raised when it is desired to **clean up** or remove the gold and as readily replaced. They are not as good for fine as for coarse gold, and if two consecutive riffles are placed so the spaces between the boards are in line, the gold may travel over the boards instead of going between them.

Slot Riffles. — Wooden riffles constructed as shown in Fig. 18, are termed slot· riffles, the slots being about 2 inches wide, 8 inches long and ¾ inch deep. Mercury

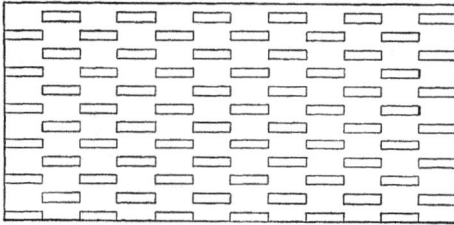

Fig. 18.

is placed in each slot. These riffles can be arranged so that the longer dimensions of the slots come either across or lengthwise of the sluice. The slope for such riffles should not exceed 8 inches in 12 feet, otherwise fine gold will wash over the mercury. In all riffles intended to save fine gold, the mercury used should be charged with at least some gold as mercury containing amalgam is better in holding fine gold than pure mercury. Riffles should never be charged full, as they are subject to a loss of mercury under the most favorable conditions, and the loss will be increased if the riffles are kept too full. As sand moves over a sluice bottom, and comes to the mercury it passes over the mercury owing to the latters specific gravity. Fine gold will also be moved over the mercury unless it is in proper physical condition for amalgamation, and it may require several riffles before the gold is finally captured. The action of mercury is not to absorb gold and form amalgam at once, but to gradually dissolve it; therefore, float gold, and what is termed spongy gold, is not easily caught by mercury on account of its lightness or a coating of some kind of material. The specific gravity of mercury being at 60° Fahr. 13.58, and native gold 19.3, or if containing silver 15.6 to 19.3, it follows that the gold will sink into the mercury bath, while sand, with a specific gravity of 2.63 to 3, will not. But mercury is not necessary to catch the heavier particles of gold, which would lodge anyway, but is useful in saving the fine gold, if it can be held in contact with the mercury a sufficient time to allow it to be dissolved.

After the formation of amalgam, which is brittle compared with mercury, according to the amount of material it has absorbed, there is danger of loss by its floating away, and this means a loss of mercury and gold.

Frequently free mercury escapes from the riffles and is lost, owing to it being subjected to shock from stones that roll along the floor and splash into the mercury trap. To prevent this the trap orifices should be made narrow so that large stones can not enter, and small stones can not enter at high speed.   Stones less than one half inch in diameter have spattered fresh mercury, when allowed to splash into riffles. The kind of riffle to adopt will depend upon the size of the operation and those so far mentioned would not be suitable for a large hydraulic proposition. Mr. A. J. Bowie considers that where coarse material is washed "**block riffles**" have advantages over any other.

    1. Because they make a cross-riffle.

    2. They are inexpensive and durable.

    3. They are convenient to tear up, clean, and replace.

The blocks for riffles may be square or round, and from 8 to 13 inches high. The squared blocks are placed in rows across the bottom of the sluice and separated transversely by strips of 1-inch boards, to furnish a mercury trap and hold the blocks in position. The strips are nailed to the blocks, and the blocks are wedged to the sides of the sluice box as an additional precaution to prevent their moving. The longitudinal fibers of the blocks are placed upwards in order that wear and tear will be lessened, and the fibers may assist it arresting the movement of the gold. The blocks are also arranged so that they will not have their joints in the same line, or are laid as a mason lays bricks by breaking joints.

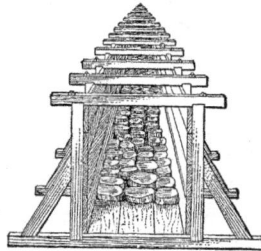

FIG. 19.

Fig. 19 shows a sluice box with round block riffles. This style of riffle is as effective in saving gold as the square block riffle, and is also used where the material washed is coarse. The blocks should be of hard wood, and held in place by strips of boards nailed to them or to the floor. In some cases cobble stones are used at the head of a sluice box where the impact of the material is the greatest, as blocks wear much faster in this place than in the sluice proper. The objection to round blocks is, that they are difficult to obtain of the same diameters and are therefore difficult to lay and fasten in the sluice, however this objection is not sufficient to prevent their use.

Stone Riffles have been used where coarse material was to be washed. Stones over the size of ones fist, and from that to stones weighing 100 pounds are considered coarse material. In large operations there is no time to assort the material and everything that the water brings to the sluice is passed through. While this is not good practice, often times it can not be avoided, particularly in hydraulicking. Sluices to handle material of this kind must be lined on the sides and floor, as in Fig. 14, to prevent their being destroyed before the season is over.

Riffles for saving fine gold are constructed so as to conform to the ideas of the operator, however his ideas are frequently modified to conform with the locality and means at command. If sluices are long, and riffles and amalgam traps are used, it may be necessary to patrol the line for the purpose of keeping watch over the riffles, and seeing that they do not become choked, do not leak, and finally that amalgam is not stolen.

Iron Riffles. — Iron rails are used to some extent as riffles. When so used they are laid lengthwise of the sluice with the flange either up or down. They are fastened together in such a way that the ends will not curl up, and they are also spaced with blocks between them.

The objections to rail riffles are, their great weight, opportunities they offer to gold to ride the rail flanges, and their cost. There are conditions, no doubt, that would favor such riffles, for instance where the rocks are not water rounded, and where there is excessive wear on block riffles. A riffle used to some extent in the Seward Peninsula [1] is shown in Fig. 20. It is a light iron casting that can be readily handled. The slots are placed either longitudinally or transversely, although the longitudinal position relative to the sluice is considered to be most effective.

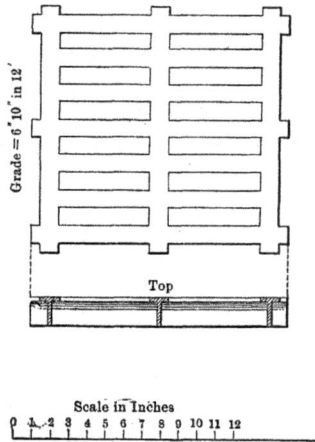

FIG. 20.

The subject of iron riffles would be incomplete without a description of the Risdon Iron Company's patent riffle, a section of which is shown in Fig. 21.

The object of such riffles is to create dead water under them and save such fine gold as mercury will not readily hold, and rusty gold that mercury will not attack. Incidentally, in accomplishing this purpose they do away

[1] C. W. Purington, *Mining Magazine*, February, 1905.

with the use of mercury, and hence loss of quicksilver
and amalgam; further, they are more easily handled and
cleaned up than the ordinary riffle. The amalgam

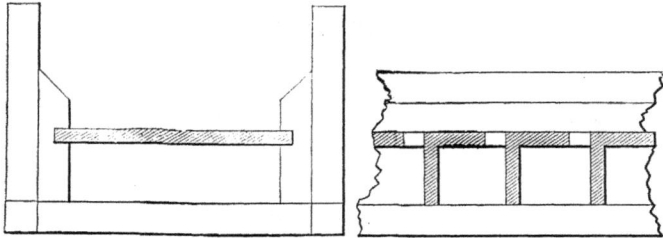

Fig. 21.

retort is thereby abolished, and the gold is recovered
purer, and commands a better price.

The riffles are made of angle iron in sections, for any
width of sluice desired. Each section is 2 feet in length,
so that the sections can be readily removed for cleaning
up. The angle irons are fastened at each side to the
box, and are spaced so that any gold passing down the
sluice along the bottom may fall into the spaces thus
created. As no water comes in except from the openings or spaces, the water under the riffles is dead, allowing
fine gold to settle and remain in the trap until removed
at "clean-up."

Float Gold.—Float gold is either in thin scales or in
such small light particles that it is termed flour gold. Mercury will amalgamate such gold if the mercury and gold
are in proper condition for alloying. If the mercury is
sickened by impurities in the dirt such as arsenic, sul-
phide of antimony, manganese, etc., it acts so sluggishly

that fine gold will move over it. Float gold will be buoyed up by muddy water, particularly water containing much clay, or talc.

Talc seems to form a sort of scum that prevents the

FIG. 22.

mercury from attacking the gold until it has been washed off. Clay acts similarly, particularly if it contains much iron oxide. Spongy gold is fine gold containing pores into which clay or other material has filtered to such an extent as to greatly decrease its specific gravity, and such gold will float over mercury, where solid grains will sink through.

Undercurrents are introduced in sluice lines to relieve

the main sluice of coarse material and save fine gold. For this purpose a grizzly made up of iron bars, set on edge, one inch apart lengthwise of the sluice, is used for a sluice bottom, see Fig. 22.

The finer material passes through the bars, while the coarser material remains on the bars. Below the bars is a coarse iron screen which checks the momentum of the coarse material and affords any gold that passes over the bars an opportunity to reach the undercurrent. The undercurrent is a shallow wooden box, from four to ten times the width of the sluice and high enough to contain the material washed into it. It is paved with either wood or stone in such a manner as to stand wear and serve as riffles. The water and material that flows swiftly in the sluice is suddenly spread over a very much larger area and this gives the gold an opportunity to settle. After the material enters the undercurrent it is spread over the entire box width by the riffles, although the inclination of the undercurrent is considerably more than that of the sluice.

The undercurrent is gradually narrowed towards the discharge end, to conform with the width of the sluice into which it discharges. In some cases the large stones left on the grizzly are not permitted to enter the sluice again, but this is not always practicable on account of dumping ground. If the water is to transport the large stones the entire quantity can not enter the undercurrent, and sufficient must pass the grizzly to carry the stones down the sluice after they are removed from the screen bars.

Undercurrents are at times great gold savers, for

example with a sluice 5 feet wide, and an undercurrent
20 feet wide, there was a saving of 20 per cent of the
gold, and upon making the undercurrent 30 feet wide an
additional 7 per cent was saved. This was accomplished
without increasing the length of the grizzly, and shows
the advantage of suddenly decreasing the velocity and
depth of the current. There may be several undercurrents in a sluice line, depending on the quantity of
fine gold and the clay in the dirt washed. The grade
given undercurrents varies from 12 inches in 12 feet, to
18 inches in 12 feet. A short undercurrent 20 feet in
length should have a steeper grade than one 40 feet long,
the reason being that the material will flow in a thinner
stream on a steep grade. The riffles in the undercurrent
will exert considerable influence on the grade, and one
can only determine their action after experimenting.
Fig. 23 is an undercurrent used on a dredge in Atlin,
B. C.[1] The grizzly has ⅛-inch spaces between the bars,
a practice to be followed where the table is but 11 feet
7 inches long.

The method of distributing the material passing
through bars as well as the mercury traps is shown
in section. The practice of using amalgam plates in
such cases is questionable, since they will be scoured in
all probability even should a large proportion of the
material be less than ⅛-inch diameter.

This undercurrent is shown here as an explanation of
the Hungarian riffle.

Hungarian Riffles. The riffle shown in Fig. 23 consists of a series of gouges made in a 2-inch plank, so

[1] Report of Minister of Mines, 1904.

staggered that they will cover the width of the sluice
bottom. This form of riffle is a favorite on dredges and
in some small sluice mines; it is not however superior
in any way to other riffles, and must not be used where
there are coarse stones, if the operator is anxious to save
mercury. Mercury is usually placed in each depression.
A somewhat similar riffle is made by boring 2-inch holes,
½-inch deep in 2-inch planks, and staggering them so
that no part of the current shall escape passing over
some of the holes. Mercury is usually placed in these
holes to catch and retain the gold. Stones lodge in
them and as they are washed out by other stones strik-
ing them there is generally a loss of quicksilver and
amalgam.

The Dump is one of the requisites of a sluicing proposition. The lack of dumping ground is often a hin-
drance to hydraulic mining and in California it prevents
many places from being worked. In the early days of
hydraulicking thousands of tons of earth were washed
daily into rivers that became clogged and changed their
channels. The material broken down occupied a larger
space than in the original bank, and spread over the
valleys, particularly the valley of the Sacramento River
in California. Fertile farm land was destroyed, making
it necessary for the government to stop hydraulicking,
until some method could be devised to conserve the farm
lands and at the same time permit mining. The construction of "**debris dams**" by appropriations from the
government and state has only partially restored the
industry. Where there is a large river into which the
debris may be sluiced by gravity without damaging

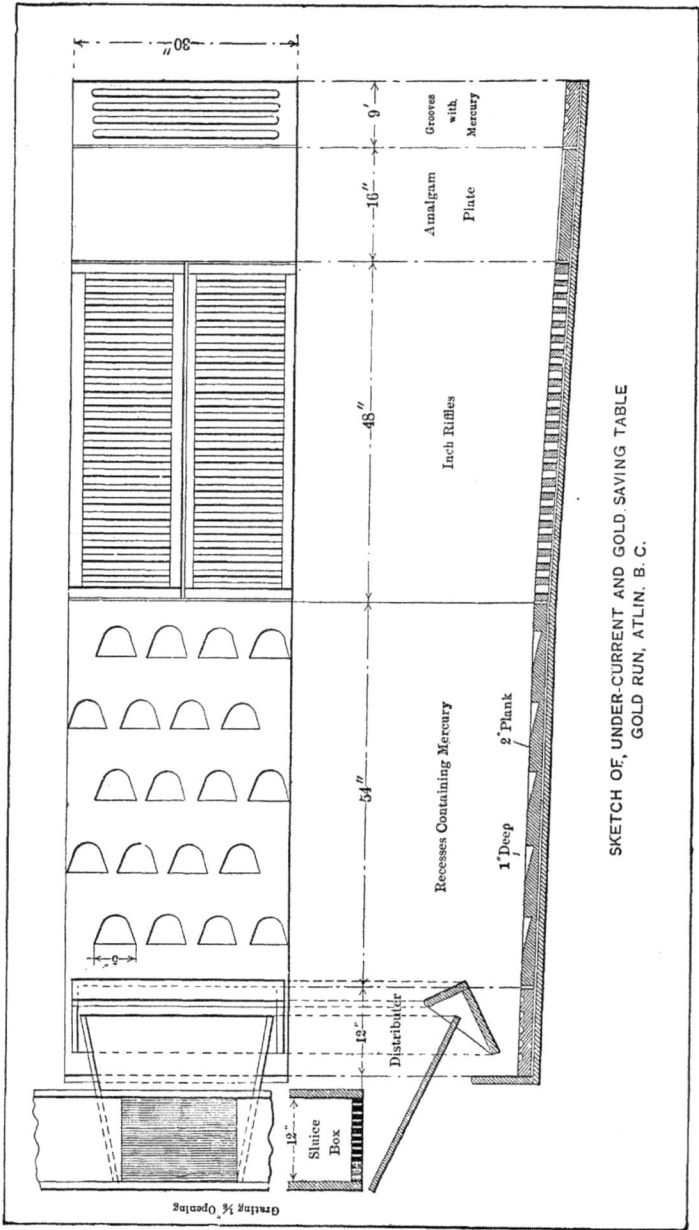

SKETCH OF, UNDER-CURRENT AND GOLD SAVING TABLE
GOLD RUN, ATLIN. B. C.

FIG. 23.

78

farm lands, hydraulicking and sluicing is still carried on, not only in California but in other localities.

The lack of dumping ground for tailings will often necessitate a lengthening of the sluice.

In Fig. 24 are shown a number of tributaries to the original sluice. These provide a wider area for the disposal of the tailings, and were necessary owing to the flatness of the dumping ground.

In some cases, as for instance at Breckenridge, Colorado, it is possible to use hydraulic elevators and thus obtain a fall sufficient to sluice the waste material to a

FIG. 24.

suitable dumping ground. To use an hydraulic elevator there is needed a large supply of water. This is lacking in some cases, and then to dispose of the debris other methods are adopted. The plan adopted at

placers where both water for hydraulicking and fall for
dumping ground was lacking is illustrated in Figs. 50
and 63.   In cases where water for elevators was lacking,
a good line of elevator buckets might be found serviceable in disposing of tailing.   They would require power,
and this might be furnished from the sluice.

CHAPTER V.

WATER SUPPLY.

WHERE placer mining operations are to be carried on by hydraulicking, the most important factor to be determined is the quantity of water that can be depended upon. In some cases water has been conducted through ditches, flumes, pipes, and tunnels for 50 miles and in one case 100 miles. This of course requires an immense capital and a thorough survey of all the watersheds throughout the length of the ditch. If the ditch can be connected with a large river or lake without too great expense, the placer miner will have an ideal water supply. This however is possible only occasionally, and for the most part the supply for placer operations must be obtained from streams supplied by melting snows and rains. While this supply may be in excess of the miner's needs at certain seasons it may be so scant in the summer months that operations must cease. It is necessary in order to ascertain what definite supply can be depended upon the season through, to examine the records of snow and rain fall, and to locate places where reservoirs may be established as feeders in times of dry weather.

To carry the survey out properly, reliable data must be obtained in regard to the average flow from creeks and springs and the area drained by them. In selecting

81

the site for a **storage reservoir** the following information
is to be obtained.

1. The elevation above the mine, so that a sufficient
pressure will be assured for operating the giants and
elevators.

2. The watershed feeding the reservoir, and the
water that may be depended upon.

3. The formation and character of the ground with
reference to the absorption and leakage that might occur.

Absorption and Evaporation. — The most desirable
ground for a reservoir site is one of compact rock, like
granite, gneiss, or slate. Porous rocks, like sandstone
and limestone, are not so desirable, on account of their
absorptive qualities. Steep, denuded slopes are best
watersheds, as then but little water will sink into the
ground and the remainder will go into the reservoir.
The longest slope will furnish the largest available
quantity of water provided vegetation does not cause
too much absorption. Bowie states that at the Bowman
reservoir, in California, 75 per cent of the total rainfall
and snowfall (reduced to rain) is stored.

A reservoir should hold a supply capable of meeting
the maximum demands. The area of the reservoir is
determined by surveys and a table made showing its
contents for every foot of depth, so that the amount of
water available can always be known. The Bowman
reservior contains about 1,050,000,000 cubic feet of
water. The catchment area or watershed is 28.94
square miles. The cost of the reservoir and dams was
$246,707.51. Beside the main reservoir, all mines
should have auxiliary reservoirs which although com-

"U-she" Dam
State Creek

DEWEY ENG.CO.S.F.

Debris Dam

paratively small are adapted for short runs. These are for the sake of insuring a supply in case of accidents to any part of the main supply ditch above them. An allowance must be made for leakage and evaporation, in the ditch line, and this loss in cubic feet per second per mile may be approximately estimated from the formula.

$$\frac{Ma}{v \times 5280}$$

in which M is a coefficient that varies from 3 to 20 according to the climate.

Storage reservoirs are particularly necessary where the water supply is from mountain streams which have a tendency to slack off in water during the summer months. The erection of retaining dams for such reservoirs is part of the ditch system.

The primary object of dams is to retain water; they therefore should be water tight. Dams must have firm foundations to prevent their sinking, and have their bases sufficiently wide to prevent their being moved down stream by the pressure of the water against them.

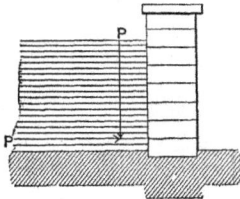

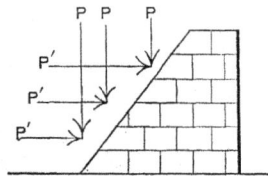

FIG. 25.   FIG. 26.

It will be necessary to increase the base of a dam in width as the dam increases in height.

Fig. 25 shows an incorrect method of building a dam wherever the volume of water is variable. The pressure, P, of the water increases with depth, and exerts a pressure, P', which tends to slide the dam off its base. If constructed as in Fig. 26, the dam will be more stable and resist the water pressure at its base, for the weight now acts in part to keep the dam in position, consequently is opposed to the pressure, P', which acts to push the dam outward.

Masonry dams are expensive, but masonry is necessary at least at the sides of any reservoir which is to contain any amount of water. The center may be crib-work, weighted down with stones, puddled clay, etc.

Crib Dams. — The crib dam, Fig. 27, is made of logs, bolted and spiked together. The ties, P, are

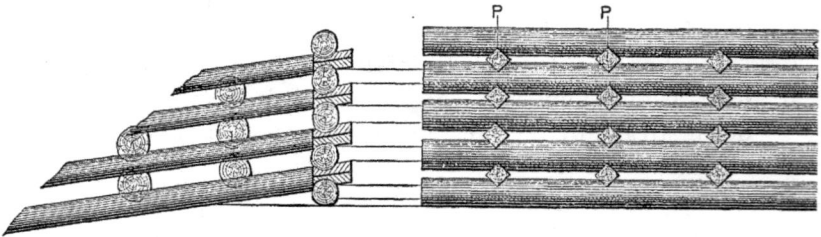

FIG. 27.

notched in diamond-shape, with a section of the log forming a collar. They are longest at the bottom of the crib, to be weighted down; they are also spiked to the log below them through the collar.

The face logs are notched to receive the diamond-shaped collar of the ties. The face logs should have

Storage Reservoir.

Retaining Dam.

the joints broken, and the ties should all be one above
the other.   This structure may be given a batter on the
outside or be reinforced by an embankment of stone.
The weighting down of the ties should proceed with
the building up.   Care should be used to puddle the
structure to prevent leakage.   Large stones if laid with
some system next to the face inside the crib will prolong
its life considerably.   The ties will not rot fast, and
the face will last many years, even when rotted considerably, if such a system be followed.

Miner's Inch. — The miner's inch, up to the year
1905, was very confusing, as each ditch company in
California at least had a water inch of its own.   The
miner's inch in California is now 1.5 cubic feet per
minute, or 11.25 gallons per minute.   For calculations
and reference the following table will be found useful.

A flow of one miner's inch of water is equal to the supply of —

	Gallons.	Cubic Feet.
Per second	.1875	.025
Per minute	11.25	1.5
Per hour	675.	90.00
Per day	16200.	2160.

Or, a flow of one cubic foot,
    Per second equals 40 miner's inches;
    Per minute equals ⅔ miner's inch;
    Per hour equals .0110 + miner's inch.

The most common measurement for a miner's inch is
through an aperture 2 inches high and whatever length
is required, over or through a plank 1¼ inches thick, as
shown in the Fig. 28.   The lower edge of the aperture
is 2 inches above the bottom of the measuring box, and

the top plank 5 inches above the aperture, thus at the center of the stream flowing out there is a 6-inch pressure of water. Each square inch of this opening will discharge 1½ cubic feet, or 11¼ gallons of water per minute, a quantity that represents a miner's inch. If the slide be moved out 1 inch the aperture for discharge will

Fig. 28.

be 2 square inches and the flow of water 3 cubic feet per minute, or 2 miner's inches.

Weir Measurement. — To form a weir and measure a small stream place a board or plank notched, as shown in Fig. 29, at some point in a stream, where it will dam the water and form a pond above it. The notch in the plank should be twice the depth for a small quantity of water and longer in proportion if a large quantity of water is to be measured.

The edges of the notch should be beveled toward the intake side, as shown. The overfall below the notch should not be less than twice its depth; that is, if the notch is 6 inches deep the overfall should be 12 inches. In the pond, about three feet or more above the weir, drive a stake, and then partially obstruct the water until it rises to the bottom of the notch, and mark the stake at this level. Then complete the weir so that all water in stream will go over the notch, and make another mark at this level on the stake. The distance between the marks on the stake, measured in inches, is the theoretical depth of flow.

FIG. 29.

To find the discharge over a weir of this description in cubic feet per second:[1]

Let h = head in feet.

        b = the length of the overfall in feet.

        c = constant number 3.33.

        Q = discharge in cubic feet per second.

Then $Q = \sqrt{h^3} + b + 3.33.$

Example. — How many cubic feet per second will flow over a weir 4 feet long, 0.64 feet deep, measured as at h or on the stake, with the constant number 3.33?

Solution. — $Q = \sqrt{.64 + .64 + .64} = \sqrt{.262144} =$ .512 $\times$ 4 $\times$ 3.33 = 6.82 cubic feet per second, or 51.15 gallons per second.

To facilitate matters for the engineer, tables of weir

[1] Trautwine.

measurements have been made. The table on page 338 gives the cubic feet of water per minute which will flow over a weir 1 inch wide and from $\frac{1}{2}$ to 20$\frac{7}{8}$ inches deep. For example, suppose the weir to be 60 inches long and the depth of the water on it to be 6$\frac{3}{8}$ inches. Follow down the column marked inches on the left until 6 is reached; follow across the table on the line with 6 until $\frac{3}{8}$ is reached, when 6.44 is found. Multiply this latter number by 60, which gives 386.40, the number of cubic feet passing per minute.

Stream Measurement. — Weirs are only adapted to the measurement of water flowing through brooks, hence larger streams are measured in some other way. To measure approximately a stream by the current and cross-section:

Measure the depth of the water at from 6 to 12 points across the stream, at equal distances apart. Add these depths in feet together and divide by the number of measurements made to obtain the average depth of the stream, and this quotient multiplied by the width of the stream will give the depth of its average cross-section.

The velocity of the stream is now found by measuring 100 feet along the bank; marking both ends of the line, and throwing a float into the stream a short distance above the upper mark. The time consumed by the float in passing the distance of 100 feet is the recorded velocity. This operation should be repeated several times, in order to determine the average velocity of the current.

One-half dozen floats thrown into the stream together and timed from the first one passing the upstream mark to the first one passing the goal will give a closer average time.

Dividing this distance by the average time found for covering it gives the velocity in feet per minute at the surface of the stream. The surface water moves swifter than the water at the bottom or sides of the stream, the difference being about 8 per cent, but for approximate calculations this need not be considered.

Flumes. — Where the life of a mine is not more than ten years, and timber is cheap, flumes may be adopted to advantage for conducting the water from the reservoir to the pressure box.

Flumes are probably cheaper to construct than ditches, and their repair is less. The calculation for the carrying capacity, etc., of flumes is the same as for sluice boxes, and need not be repeated. In most **ditch lines** flumes must be used for a portion of the distance, there being no other feasible method of overcoming certain difficulties in the construction of the ditch. It is better if possible to carry a ditch around the head of a cañon, than to cross the cañon with a flume on a trestle or siphon it through pipes. While as a rule it is better to avoid flumes, there are situations where they can not be avoided, in fact a flume was bracketed to a cliff, 118 feet above the bed of a ravine and 232 feet below the top of a cliff in California.

Where rock is to be excavated, or where porous ground is met, or where chasms are to be crossed, recourse must be had to the box flume.

In building a flume, nothing smaller than $1\frac{1}{2}$-inch plank, tongued and grooved, should be used, and these joints should be white-leaded and well driven home. The sides should be made of the same material, dry pine or spruce — the latter is preferable — and be thoroughly fastened to the posts. The sills should not be over 4 feet apart, should project 18 inches beyond the outside of the box to take braces, and in cases where there are tunnels or trestles they should project far enough to receive a 12-inch plank for a walk, in order that the flume may be examined; in other situations the plank placed along the cap pieces will be sufficient. In laying the lining, care should be taken to break joints and the lining should be first-class lumber, free from knots.

The grade given the flume will have some bearing on its size — the steeper the grade the more water the flume will pass for a given area; this will allow considerable decrease in area over the area of the ditch, and consequent economy, wherever the whole grade from the source to the outlet may permit of an increase. The flume grade may be increased from ditch grade of 0.25 per cent or 13.2 feet per mile to 0.50 or 0.75 per cent.

It is not always customary to employ side braces for the flume; they should, however, be used in certain situations. Wherever the flume crosses a ravine on trestles, braces will make the whole structure more rigid against wind, even when the trestles are anchored by wire ropes; again, on the side of a hill or cliff, where the flume runs full at one time and half full at another, braces will tend in a measure to prevent warping, especially wherever the sun's rays strike the flume.

In this connection it is well to allow a little water to run over the bottom of a flume at all times, to keep the

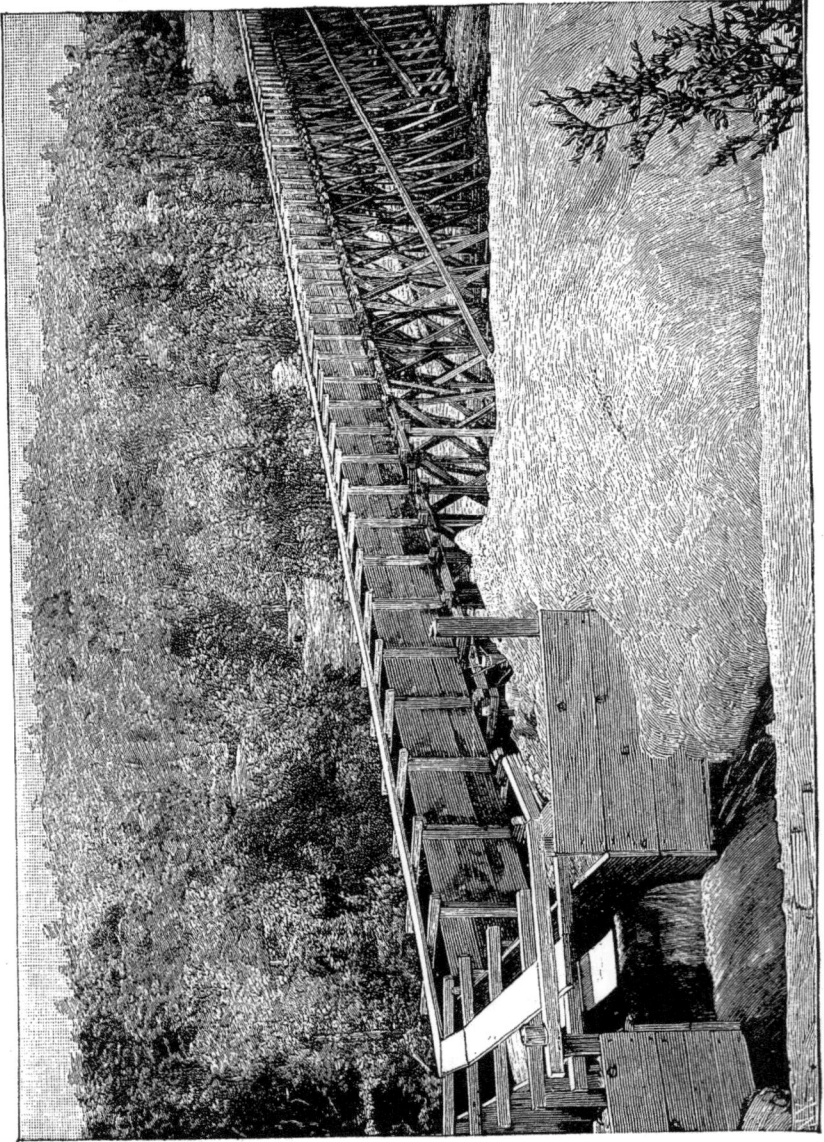

FLUME ACROSS GULCH.

joints tight, as the change from dry to wet conditions invariably causes leakage.

The size of a flume will decide the timber to be used in its construction — that is, a flume 2 × 1½ feet will not require as heavy lumber as one 3 × 3 feet in sectional area, except for lining.

The sills, posts, and cap pieces of a flume should not be over four feet from center to center, and if three feet between centers it will be more rigid.

In building flumes it is not altogether what they bear in weight, for there are other factors, such as leakage and warping, which must be guarded against, and with sills far apart the latter material elements in the problem have greater play. In the construction of a flume 3 × 3 feet the following timber would be required for 100 feet, the sills, caps, and posts being placed three feet between centers:

34 sills, 3 × 4 × 6′ 11″	=	235	feet,	2 inches.
68 posts, 3 × 4 × 3′ 2″	=	215	"	4 "
34 caps, 3 × 4 × 4′ 3″	=	144	"	6 "
68 braces, 3 × 2 × 3′	=	102	"	0 "
54 lining plank, 12 × 1½ × 15′	=	1215	"	0 "
9 lining plank, 12 × 1½ × 10′	=	135	"	0 "

Per hundred feet,   2047 feet, 0 inches.

Per mile, 52.8 times,   108,082.

Note. — The above bill does not include walk or battens; add 11,780 feet for 1 × 3″ battens and 19,720 feet for 1½ × 12″ walking plank per mile.

The lumber per mile of flume does not include stringers

or blocks, which must be placed lengthwise of the flume on trestles; in this connection it must be borne in mind that the foundation for a flume must be solid and level, especially under each sill. The usual size for stringers in a 3 × 3-foot flume is 4 × 6 inches, but this size must be determined by the distance between bents in the trestle, since the stringers, as well as supporting the weight of the flume and water, tie the trestle bents. With bents 12 feet between centers the stringers should be 6 × 12 inches for this area of flume.

The sills should be notched for the posts; the caps should be mortised for tenent at the top of the posts, and secured by $\frac{1}{4}$-inch wooden pins. The general practice is to make notches for both caps and sills, using spikes to hold them.

Where curves are necessary in the flume the outer side of the flume must be raised, to correspond to the degree of curvature; $\frac{1}{2}$-inch elevation from the lower side for every degree of curvature will be sufficient. This elevation should commence on the straight line or tangent of the ditch before it meets the curve, as this will tend to equalize the flow. The elevation should be gradual and reach its height at the center of the curve, and as gradually recede, until the flume again becomes straight, the object being to change the motion with the least friction possible and avoid the water pouring over the side of the flume at the center of the curve. Whereever curves are met the sills and posts are set closer, and greater care is to be observed in placing the lining.

At times it becomes necessary to run along the side of a cliff; this is accomplished by drilling holes in the

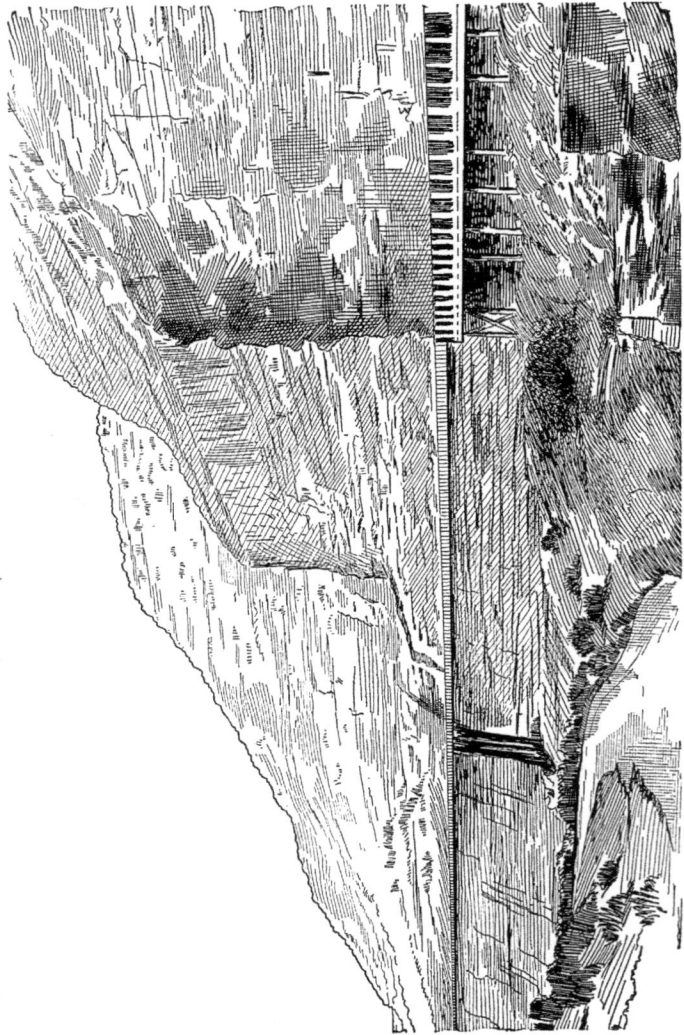

Cliffside Flume.

cliff and putting in iron brackets, upon which the stringers for the flume rest. The brackets curve upward
parallel to the posts, and are fastened by anchors to the
cliff above the flume. In San Juan County, Colorado,
flumes are carried some distance in this manner. Flumes
should have 3 × 1-inch pine battens over the floor-
seams to prevent wear, and should be provided with
gates at intervals, to allow water to be drawn off; further,
where snow or dirt is likely to slide into them from the
mountain side, they should be covered with sheds.

A common method of constructing flumes, and placing
them on half-bents is shown in Fig. 30.

This method was probably introduced first by the
North Bloomfield Mining Company in California, in

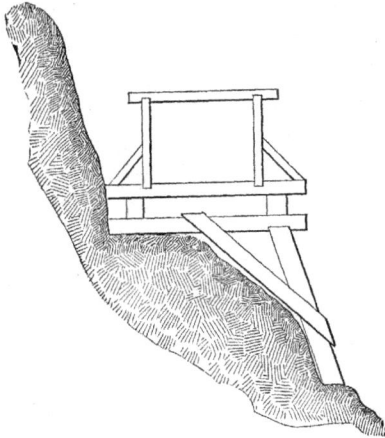

FIG. 30.

their ditch line between Eureka and Milton dam. The
slope of the rock in some places was such that to con-

tinue the ditch would call for an enormous outlay of money, and as it was, the foundations for the 5.3 miles of flumes cost $18,920. It will be observed that the posts are dapted into the sills and caps of the flume, and that the stringers are also dapted for the bent cap. For a 3 × 4-foot flume, lined with 1½-inch planks and battened where the planks join the caps and. legs, are 8 × 8-inch timbers, and the braces are 3 × 8-inch timbers. The cap must rest firmly on the foundation, and a proper faced hitch must be cut for the leg. In addition to these precautions holes are drilled in the rock and the timbers spiked to the rock.

The stringers are 8 × 10-inch timbers dapted for the bent sill. Stringers should be placed over the bent legs, and under the flume posts, as the weight will then be transmitted properly to the ground.

The sills and posts are of ∴ × 5-inch timbers.

The posts are of 4 × 5-inch timbers, and the sills of 4 × 6-inch timbers because of the notches cut. The caps are of 3 × 4-inch scantling, and the braces are 2 × 3-inch planks.

Flumes are often placed on trestles to cross narrow gulches, some of which are from 25 to 50 feet high. Fig. 31 shows the method of constructing such flume-bents where the height does not exceed 20 feet. The dimensions of the flume for such situations are about as stated, as there is no necessity of increasing their depth or width if the proper grade is continued over the trestle. The trestle timbers in all cases should be calculated for the load they are to carry, and an engineer employed for their design and construction. In most

ditch lines there is too much money involved to make
any part weak, and as one part is dependent upon the
other, should a break-down occur, the entire system is
stopped. It is customary to
arrange overflows at stated
intervals along the ditch lines,
to let any surplus water arising from rains, snows, or
freshets escape. These are
placed about three-quarters
of a mile apart, and at feeder
stations or auxiliary storage reservoirs. When ditches
have become clogged with

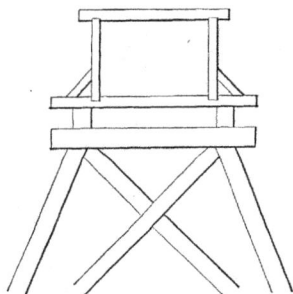

FIG. 31.

snow after a shut-down they must be cleaned out
before the water is turned in, otherwise time will be
consumed, much trouble encountered, and possibly
damage. Water will not melt snow fast, and will not
flow through it at all, for which reason sections of the
ditch must be shoveled out or washed out. In the
latter case a breach is made in the ditch and the water
will then float the snow through it rapidly. When
cleared of snow the ditch is repaired, and a similar
breach made lower down the line, and so on to the
flume. Long flumes must be shoveled out, or at
least a channel made through the snow in them.

CHAPTER VI.

IN 1852 Edward E. Mattison, from Connecticut, with a view of economizing labor, in California placer mining conveyed water to his claim in a rawhide hose to which was attached a wooden nozzle, for spurting the stream against the gravel bank. This was the first step in modern hydraulic mining, and was so appreciated that canvas hose bound with wire and rope soon followed, and the nozzle was changed from wood to metal.

The canvas hose was soon superseded by the invention of R. R. Craig, who used at American Hill, Nevada County, California, about 100 feet of stovepipe. A firm in San Francisco, according to A. J. Bowie, commenced the manufacture of wrought-iron pipe for hydraulic mining in 1856. The great difficulty experienced with such pipes was the quickness with which they rusted. They were, therefore, painted on the outside, but this did not prevent their rusting on the inside.

As pressure became an item of importance, the strength of the pipe was also a consideration, and as iron pipe was costly and difficult to transport, attention was given to wrought and sheet-steel pipe made in lengths suitable for transporting on mules or burros. Spiral-riveted, galvanized iron pipe was first introduced, but this gave way to riveted sheet-iron and sheet-steel pipe, that is made in sizes from 4 to 60 inches in diameter, and is capable of resisting pressures up to 600 lbs per square inch. The general impression prevails that such pipe is not suitable either for pressure or permanency, yet the Connecticut Tube Works have been making for municipal service a sheet-iron pipe lined with cement for some years, which they claim is more serviceable than cast iron and fully as strong after fifteen years' service.

Properly constructed sheet-metal pipe, when painted with asphalt inside and out, to prevent corrosion, has lasted twenty-five years and come into general use for hydraulic mining. The numerous changed conditions to which this kind of pipe has been subjected have furnished reliable data in regard to pressure, diameter, and thickness of metal required for various pressures.

The result of this experience, briefly stated, is, that a comparatively light sheet-metal pipe, in sizes properly proportioned to diameter and pressure, is both cheaper and more satisfactory than any other pipe for hydraulic mining.

Asphalt paint, so long as it is kept intact, makes the pipe practically indestructible so far as ordinary wear is concerned. Where the coating is worn off by abrasion in transportation, or where the pipe is subject to severe shock by the pressure of air [1] on suddenly closing the gates, or where expansion and contraction open the joints and break the asphalt, corrosion would naturally occur, but this can be remedied by care and an application of paint to such places.

In laying pipe the shortest practicable distance is

[1] Water hammering.

advisable, wherever the ground will permit it, and sheet pipe should always have a solid foundation along its entire length. If it must cross a small ravine it should be on a trestle with its entire length resting on plank. Short turns or acute angles should be avoided, as they lessen the pressure and strain the pipe; also, the pipe will be more affected by expansion and contraction at such points.

Wherever practicable, the pipe should be laid in a trench and covered with earth, to protect it as much as possible from contraction and expansion or injury. When laid over a rocky surface, straw or rubbish will protect it from the sun, and generally prevent freezing, especially, if the water is in motion. As a rule, pipe is not often used along the ditch line, but runs from the reservoir at the end of the flume termed **pressure box** down a steep incline to the mine.

Pipe laying should commence at the lower or discharge end and proceed up the hill. In the long-distance transmission power plant at Fresno, California[1] the construction of the pipe line commenced at both ends, and considerable difficulty was encountered in closing the gap at the center of the line. This was due to the alteration in length resulting from the change of temperature. Before sunrise the opening would be 7 feet 8 inches, but in the afternoon the gap would be 7 feet. The connection was finally made before sunrise, and the pipe filled with water before the sun had a chance to expand it.

There are two methods of joining pipe lengths, as shown in Fig. 32. With the slip joint the pipes are not

[1] *Scientific American*, March 27, 1897.

Fresno Power Plant.

of large diameter or under very high head, and whenever this stovepipe joint is used, the lower end of each
length of pipe is wrapped with cotton drilling or burlap,
to prevent leaking, inserted into the next lower length,
and driven in. Where slip-joint pipe is to be used an
allowance of three inches must be made on each length
of pipe ordered, for loss in driving the joints together.
In case they leak but slightly, the leak may be stopped
by throwing bran or sawdust into the pipe; or if that
does not answer, dry wooden wedges are to be driven
into the joints. Should the leak be large, clamps must
be used which encircle the joint.

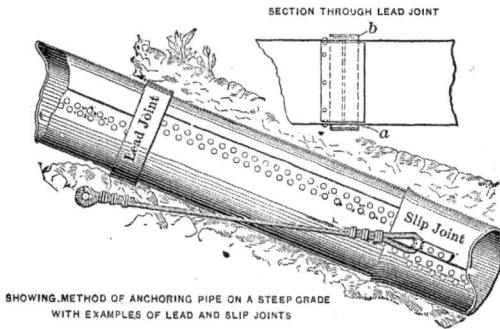

SHOWING METHOD OF ANCHORING PIPE ON A STEEP GRADE
WITH EXAMPLES OF LEAD AND SLIP JOINTS

FIG. 33.   FIG. 32.

In laying pipes where the lengths come together at
an angle a lead joint, Fig. 33, should be used, or where
the pressure is great or the diameter of the pipe is large
lead joints should be made. This joint is made by
means of a sleeve, a, which has a diameter $\frac{3}{4}$ inch larger
than the pipe and into the space, b, hot lead is poured.

With heavy pressure on steep grades, the pipe sections should be wired together, and lugs should be furnished on the outside of the pipe for this purpose. Anchor wires should also be attached to the pipe and to a stable object at intervals on heavy grades. It is customary to make the pipe of large diameter and of light weight metal near the pressure box, and to decrease the diameter and to use heavier metal toward the discharge.

At the Fresno power plant the pipe line was 4200 feet long, with a head of 1411 feet, giving a pressure of 609 pounds per square inch. This was built in three sections, as follows:

1st Section. — 1820 feet, 24-inch riveted pipe, first half No. 12 steel, and the second half ¼-inch steel plate.

2d Section. — 400 feet, 20-inch diameter, lock-jointed welded pipe.

3d Section. — 1800 feet, 20-inch diameter lap-welded ⅜-inch thick pipe, with flange joints and rubber packing.

This column of water weighs about 317 tons, and has a thrust of 93 tons, when issuing from a 1⅛-inch nozzle at a speed of 9000 feet per minute.

Air escaping from the Fresno pipe nozzle makes a noise which can be heard several miles. The noise is due to the expansion of air as it leaves the nozzle in bubbles that have been subjected to the heavy pressure.

The Flow of Water Through Pipes. — Head of Water. — By head of water is meant the difference in elevation between the inlet and outlet of a pipe, plus the height of the water above the center of the pipe inlet. Water pressure is due to the head, and is derived from the weight of the water; hence the higher the head, the greater will be the pressure. The pressure due to a column of water 1 inch square and 12 inches long is at ordinary temperature about .434 pound. For approximate calculations the pressure may be considered .5 pound per square inch for each foot in height.

Loss of Head. — Friction in pipes may diminish the pressure due to the head and hence the power in three ways.

1. Resistance to the flow of water is greater in small than in large pipes. The resistance does not arise directly from the rubbing surface, but is due to a layer of water that adheres to the pipe and acts as a drag on the current. The amount of drag is greater in rough than in smooth pipes, and in short bends than in long bends.

The circumference or frictional rubbing surface or perimeter of a 1-inch pipe is 3.1416 inches, while the perimeter of a 2-inch pipe is 6.2832 or twice that of a 1-inch pipe. The area of a 1-inch pipe is .7854 square inch, while the area of a 2-inch pipe is 3.1416 square inches or 4 times as great. From this it is deduced that by increasing the area of a pipe the frictional resistance is decreased, for in the 2-inch pipe with twice the rubbing surface of the 1-inch pipe, 4 times the water will pass the head being the same. The loss of head due to friction in pipes is difficult to calculate, in fact it is not necessary to make such calculations, as they have already been made and tabulated for the use of engineers. (See table on p. 339.) The formula given for friction of water in pipes is the simplification of Weisbach's formula by Coxe:

$$F = \frac{4\,v^2 l \times 5\,v - 2}{1000\,d}\ .$$

F represents the friction head or total loss by friction in feet; l the length of the pipe line in feet; d the diameter of the pipe in feet; v the velocity of the water in feet per second.

2. The flow of water through pipes depends upon the diameter and length of the pipe, and the velocity due to the head principally, but in addition there is a loss of head due to power absorbed in giving the water a uniform rate of flow or as it is termed *setting it in train.*

3. Another amount of power is consumed in overcoming the resistance due to the water entering the pipe.

A series of 88 experiments made by Hamilton Smith, Jr., on the flow of water through circular pipes of various diameters from $\frac{1}{2}$ inch to 4 feet are reduced to the formula:

$$v = m \left(\frac{dh'}{l}\right)^{\frac{1}{2}}$$

where

v = velocity in feet per second,

d = diameter of pipe

l = length,

h' = effective head,

m = variable coefficients.

The effective head h' was derived from the total

head h as follows, c being the coefficient of contraction at entrance:

$$h - h' = \frac{v}{2\ gc},$$

in which

$$g = \text{acceleration of gravity.}$$

Strength of Pipes. — Pipe lines generally involve considerable outlay, and must be proportioned for strength as well as capacity. The bursting and safe strength of iron and steel pipes, and their construction should be understood, although the table on pp. 336–338 gives the safe working pressure for double-riveted pipe up to 42 inches in diameter.

To calculate the safe working pressure for iron or steel-plate pipes, the following formula advanced by Professor Rankin may be used. In the formula P is the safe working pressure in pounds per square inch; T is the tensile strength of the plate, iron being taken at 48,000 and steel 62,000 pounds per square inch; t is the thickness of the plates in decimals of an inch; c the factor of safety usually assumed as 4; f is the proportional strength of plates after riveting, the factor being .7 for double-riveting and .5 for single riveting; R is the radius of the pipe in inches.

$$P = \frac{1}{c} \left(\frac{T \times t}{R} \right) \div f.$$

Example. — Having a head of 75 pounds per square inch, of what thickness should a double-riveted 36-inch diameter iron pipe be, with 4 used as a factor of safety?

Solution. — Factoring the equation for t, the formula becomes $t = \dfrac{cPR}{Tf}$ and by substituting the values given

$$t = \frac{4 \times 75 \times \frac{36}{2}}{48,000 \times .7} = .16 \text{ inch. \quad Ans.}$$

Example. — What will be the safe working pressure for a pipe 36 inches in diameter, made of .16-inch sheet iron and double-riveted, where using a safety factor of 4?

$$P = \left(\frac{T \times t}{cR}\right) \div f = \left(\frac{48,000 \times 16}{4 \times 36}\right) \div .7 = 75 \text{ pounds}$$

pressure. Ans.

When sheet-iron pipe is left to the option of the maker the lengths are generally 27 feet. When it is to be transported by wagon the lengths are 20 feet. When the pipe is for heavy pressure and mule packing it is made in sections of 24 to 30 inches in length, rolled lengthwise and punched, with rivets furnished to put the pipe together on the ground where laid. Sheet metal for this purpose can be riveted cold, with the ordinary riveting and flanging tools. Sheet iron or steel in this form has a discount of 30 per cent from completed pipe. After riveting, the pipe should be tarred or painted with asphalt and allowed to dry.

Pipe should be dipped in asphaltum heated to a temperature of 300° F. and allowed to remain in the bath until the metal attains the same temperature. The material for pipe construction should have No. 8, 10, 12, 14, 16, and 18 B. G., thicknesses, according to pressure, and iron plate will usually prove more satisfactory than steel plate, as the latter oxidizes and flakes readily.

Pipe Elbows. — The additional head required for bends is given on pages 314 and 318. No bends should be allowed in a pipe line, where the pipe is long or the pressure heavy, that have a radius less than five diameters of the pipe. The simplest rule for calculating the loss of head due to bends of various angles in a pipe is

$$H = .0152 \ V^2 \ K.$$

in which K is a coefficient to be taken from the following table:

Angle of Bend	20°	40°	60°	80°	90°	100°	120°
K	.046	.139	.364	.74	.98	1.26	1.86

When the radius of the bend is greater than five diameters of the pipe the loss may be calculated by multiplying the number of degrees in the angle by the square of the velocity in feet per second, and dividing the product by 88,489.

Example. — With a bend having an angle of 100° and discharging water at a velocity of 20 feet per second, the loss of head, will be

$$\frac{20^2 \times 100°}{88,489} = .45 \ \text{feet.}$$

When the radius is less than five times the diameter, fairly accurate results may be obtained, by multiplying the square of the velocity of the water in feet per second, by C, a coefficient having the following angles:

10° C. = .000109	50° C. = .003634
20° C. = .000466	60° C. = .005652
30° C. = .001134	70° C. = .008276
40° C. = .002158	80° C. = .011591
90° C. = .015248	

There has not been sufficient investigation on this subject to enable the engineer to make exact allowance

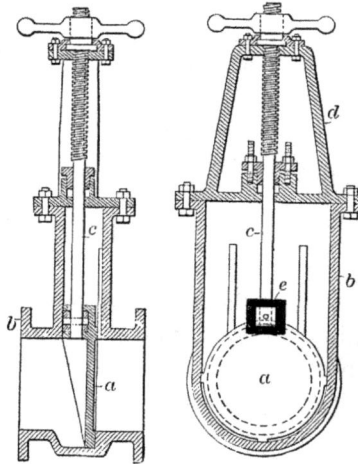

FIG. 34.

for friction due to bends, and all calculations are therefore approximate.

Water Gates. — Pipe lines would be incomplete without water gates. A section and cross section of a gate valve is shown in Fig. 34. The gate, a, slides vertically up and down, so that when fully open there is practically no interference with the flow through the pipe. The valve gate casing, *b*, is cast iron, reinforced by a web in its circular part and with a stem, *c*, terminating in a screw. By means of the movable nut provided with levers in the top of a yoke, *d*, the screw is made to turn and move

FIG. 35

the gate upwards or downwards. The lower end of the stem is fitted in a box collar, *e*, in order that it may turn freely. Whenever a junction is made with another pipe

line the custom is to fork the lines rather than use elbows.
Two such gate valves are then used, one in each branch
pipe, as shown in Fig. 35. All gate valves should have
outside yokes and coarse screw threads to prevent quick
closing and the consequent water hammer.

Air Valves. — There should be two working faces in
any hydraulic mine, so that one may be worked while the
pipe is being advanced in the other. This is accomplished by running the main pipe line into the center of
the mine and using a Y which has water gates on each
branch line. The stand pipe shown in Fig. 35 is an air
chamber, supplied with a pop valve at the top that
allows the air to escape after it reaches a certain pressure.
Its object is to prevent water hammer, and possibly
damage to the pipe line, also to collect air before it leaves
the pipe nozzle. To allow the escape of the air from a

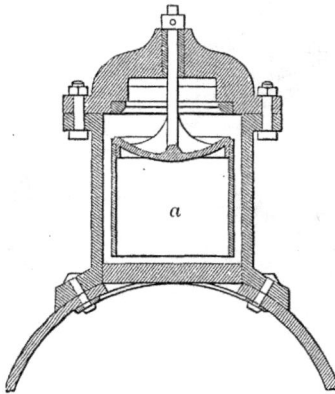

FIG. 36

pipe line while filling, and also to prevent the formation
of a vacuum and collapse of the pipe in case of a break

in the pipe line, air valves are required. The valve shown in Fig. 36 is automatic in its action, and quite simple in comparison with some. When the water fills the pipe it raises the valve, *a*, and when it leaves the pipe the valve, *a*, immediately drops and allows air to enter, thus preventing a collapse. Air valves of this description or some other should be placed wherever there is a knuckle or high place in the pipe line and where air is likely to accumulate or a vacuum occur. At all low places similar blow-off valves should be placed.

The Pressure Box. — In order to prevent sand, gravel, sticks, and rubbish from going into the pipe line, and

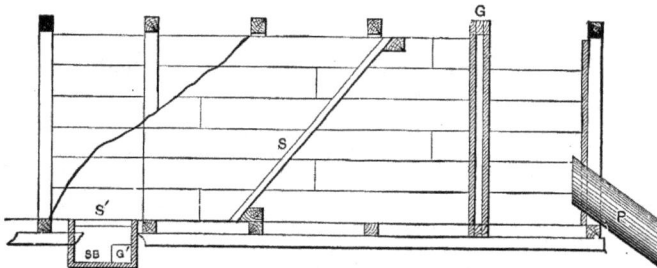

FIG. 37

particularly to prevent the admission of air, a pressure box is used. The pressure box, Fig. 37, should be large, and the water should stand at least 4 feet over the entrance to the pipe in order to prevent the admission of air. The pipe should be funnel-shaped where the water enters it, if it can be placed horizontal, but if it can not be so placed the pipe should be as in the cut, and firmly anchored.

There should be a gate, *G*, at the reservoir or flume

at the head of the pipe line to cut the water off. There should also be pressure indicators and water regulators which will regulate the flow. The cheapest gate at the head of the pipe line or along the ditches and flumes where pressure is not excessive is constructed of plank about as long as the water course is wide, and 8 inches high. These are placed one above the other, in grooves, so they may easily be removed and replaced. The grooves are formed by nailing 2 × 3-inch plank to the side of the flume, and through these guides the gate planks are lowered and raised from the top. Considerable trash is at all times moving with the water in the ditches, hence for floating rubbish the flume and pressure box should have inclined bars of wood or iron to prevent it reaching the pipe. Sand is collected by placing iron bars, S', across the bottom of the flume over a box, SB, let into the bottom as shown.

To prevent sand and gravel from entering the pressure box, there should be an increased depth and width to the ditch for at least 100 feet back from the pressure box. There should also be a small gate, G', in the sand box for blowing out, and the pipe, P, should be from $2\frac{1}{2}$ to 4 inches above the floor of the pressure box.

The gate, G, regulates the flow of water into pipe, P, and shuts it off entirely if a small waste gate is placed on the flume side. The same construction for gates may be used in dams, flumes, and ditches. Water gates are expensive when made of metal, but in some instances they will be required in the pipe, near the pressure box.

Under heavy pressure they are not easy to work, and wear out fast.

Filling Pipes. — Care must be taken when filling pipes to introduce the water gradually, in order to prevent serious accidents. For this reason it is probably a good plan to place a gate valve in the pipe below the pressure box, in order to regulate the intake flow. Air will enter a pipe in surprising quantities, and in one instance enough air was taken in at a power plant to run an engine or a rock drill. The water before it enters the pipe should be free from air and should enter quietly, this as mentioned can be accomplished to some extent by placing the pipe some distance under water, say 4 feet. Before filling the pipe all stones and rubbish are to be removed.

Ditch Lines. — Surveys are necessary to the design and construction of a ditch, as the course of the ditch is confined to narrow limits by the topography of the country through which it passes.

The survey will begin at some point very near the storage reservoir site, and will generally follow the same valley as the stream that was impounded for a considerable distance. It will be found that careful surveys in such operations pay because more accurate work can be done in their construction. As far as alignment is concerned, this survey does not call for any great degree of accuracy, the leveling being of much more importance. Errors are liable to occur in leveling. Therefore, when the alignment has been completed and leveled, check levels should be run back over the entire line. It will not be necessary, to verify the entire profile, a check on the benches being sufficient. All important tributaries should also be surveyed, carrying the survey to an elevation approximately equal to that of the reservoir site. The approximate length, together with the total fall obtained from this survey, will enable the engineer to make a preliminary design of the section and grade for the ditch.

A trial line for the ditch can now be run. For this purpose, suppose a grade of 13.2 feet to the mile is decided on for the slope of the ditch. The tangent of the angle corresponding to this slope is $\frac{13.2}{5280} = .0025$, which corresponds to an angle of nearly 9 minutes. Having a transit provided with a vertical limb, let the telescope be depressed to this angle and clamped. When the transit is set up, let the target of a leveling rod be set at the height of the telescope of the transit from the ground. This can be sufficiently approximated by holding the rod alongside of the transit and sighting across the wyes. Let the rod now be taken as far ahead as possible and moved along the ground, up or down hill, until the center of the target is bisected by the horizontal cross-hair of the transit. The foot of the rod is on the ground falling at the desired rate, and a plug should be driven at this point and the distance measured. The direction will be ascertained by the needle, as this will be sufficiently accurate. From time to time measurements will be taken to convenient stations on the line of the survey, if one has been made, as a check. It will be well to carry this line along, following all the indentations and tributary valleys, for in this way the length of a line following the natural surface of the ground for its entire distance will be obtained.

Transporting Pipes for the Line.

126

It will be very rare that this line is actually followed by the ditch. Valleys will be crossed by trestles or siphons and hills will be tunneled, but only in this way can an estimate of the comparative advantages of alternative ditch lines be compared.

When an approximate location of the ditch has thus been determined, the line will be accurately re-run and leveled over, so as to establish the final location and make a more nearly exact estimate of cost.

The material in which the ditch is excavated will place restrictions on the velocity of the water. The velocity should be sufficient to prevent the deposition of silt and not so great as to erode the bottom and sides of the ditch. The grade necessary to maintain a uniform velocity within the desired limits will depend on the interior surface of the ditch, being very much less for one having a smooth lining than for one having a rough lining. The area of cross section also is a function, for the water in a large and deep ditch will move with a greater velocity under a given grade than that in a smaller and shallower one having the same grade. The form of the cross section also exerts an influence on the velocity of flow, so that the determination of the grade becomes a complex problem, depending on the desired discharge of the ditch, its rubbing surface and form, and the dimensions of its cross section

Gravity is the sole force that acts on water in a ditch to produce the motion which takes place. The inclination of the surface of the water in the ditch is the immediate cause of motion, being that which enables gravity to act.

It is evident that the steeper the ditch grade the greater will be the velocity of the water; and as this grade is determined by the ratio of the vertical height to the distance in which it is overcome, it is evident that the accelerating force producing velocity will be expressed by the ratio $\frac{h}{l}$, in which h = the difference of level between the two extremities of the ditch and l = the distance, usually measured horizontally, separating the two.

If there were no resistance to the flow of water through the ditch, the constant accelerating force would cause the velocity to go on increasing indefinitely. Owing, however, to resistances the water soon acquires a constant velocity, provided the ratio $\frac{h}{l}$ remains constant. There are resistances that increase in intensity with the increase of velocity, so that after a certain time the increasing resistance just equals the increasing acceleration, and the velocity then becomes constant or assumes a **permanent regimen**.

The laws bearing on the subject of the flow of water in ditches, may be expressed as follows:

I. *The resistance for any given velocity is proportional to the wet perimeter or the surface over which the water flows.*

II. *This resistance affects the entire volume of water being greatest for the film in immediate contact with the wet perimeter, and becoming less and less for the films and threads more remote from that surface.*

III. *The greater the surface in contact with a given*

volume of water, the greater the resistance becomes; conversely, the greater the volume subject to a given resistance, the less will the velocity be affected.

IV. *The resistance is nearly proportional to the square of the mean velocity of flow.*

V. *The resistance varies with the nature of the ditch ground, being greater for a rough surface and less for a smooth one.*

Let

h = difference in level between ends of the ditch or any two cross sections of the ditch.

l = horizontal length of that portion of the ditch included between the sections whose difference of level is h.

g = grade = ratio $\dfrac{h}{l}$.

a = area of water cross-section.

p = wet perimeter.

r = hydraulic mean radius = ratio $\dfrac{a}{p}$.

c' = coefficient depending on the ground in which the ditch is excavated.

v = the mean velocity of flow.

Then, the resistance to flow may be expressed by the equation $ha = c'lpv^2$, from which the formula

$$v = \sqrt{\frac{h}{c'l} \times \frac{a}{p}} = \sqrt{\frac{1}{c'} \times g \times r}$$

is derived.

By replacing the factor $\sqrt{\dfrac{1}{c'}}$ by an equivalent factor, c,

then $$v = c\sqrt{rg}.$$

Form for Ditches. — It is evident from the formula
that the velocity increases with the hydraulic mean radius
$r = \dfrac{a}{p}$, and that therefore the most favorable shape of
cross section will be the one in which a given area is
enclosed by the smallest wet perimeter. In the case of
a ditch, this section would be a half circle, since the
circle is that geometrical figure which encloses the greatest
area within a given perimeter. In the case of the circle,
the value of the hydraulic radius is

$$r = \frac{a}{p} = \frac{\frac{1}{4}\pi d^2}{\pi d} = \frac{d}{4},$$

and, since both the area and the wet perimeter of a half
circle are, respectively, equal to one-half of the area and
wet perimeter of a circle when running full, the ratio $\dfrac{a}{p}$
for the half circle is also equal to $\dfrac{d}{4}$.

The half-circle form is an impracticable form for a
ditch, since it could not be constructed and maintained
unless lined with masonry or some other permanent
material, and even then the constructional difficulties
would generally render this form inadvisable, as entail-
ing a considerable expense for labor without a corresponding economy of material. An approximation to this
best form is half a regular hexagon, in which

$$r = \frac{D\sqrt{3}}{8} = \frac{\sqrt{3 D^2}}{8}.$$

D being the diameter of the circumscribing circle. This
form would also require a permanent lining if it were

applied to an earthen ditch, and would not, therefore, always be consistent with the character of the ground and the velocity of flow. The form of a ditch, therefore, should not be hexagonal, from the fact that unless the sides are of hard rock they will wash considerably. Ditches having the trapezoidal form can have their sides made to conform with the natural slope of the material and its hardness. Again trapezoidal ditches offer less rubbing surface for equal water areas than rectangular ditches. On the other hand the trapezoidal form is less adapted to withstand losses of water from percolation and evaporation that occur, owing to the large area of water surface exposed to the air, and its largest area of ground exposed to the pressure of water. The re-

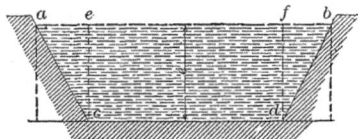

FIG. 38

lations existing between the various dimensions of a trapezoid are best illustrated graphically. The area of a trapezoid is found by adding together the length of the parallel sides, dividing the sum by 2, and multiplying the quotient by the perpendicular height ec. Thus in Fig. 38

$$\text{area} = \frac{ab + cd}{2} \times ec.$$

The angle of slope eac is equal to $\dfrac{ac}{ec} = \cot. eac$. The perimeter is $ac + cd + db$. The side ac is the <u>hypothenuse</u> of the right triangle aec, and hence $ac = \sqrt{ae^2 + ec^2}$.

The relative slope of the sides of a ditch are expressed by stating the ratio of the base ae to the height ec of the triangle aec.

Size of Ditches. — The following table is composed of subject matter obtained from Molesworth's Pocket Book" and from C. C. Longridge, "Hydraulic Mining."

It will be found useful for comparison between the different ditch sections, and for ascertaining the size of ditches to carry a given quantity of water.

Angle α Degrees.	Slope of Sides.	Vertical Depth.	Width at Top.	Width at Bottom.	Perimeter $p = \sqrt{a}$ × Factor, Factors.
90° 00′	Vertical	$.707\sqrt{a}$	$1.414\sqrt{a}$	$1.414\sqrt{a}$	2.828
78° 41′	.200	$.734\sqrt{a}$	$1.510\sqrt{a}$	$1.217\sqrt{a}$	2.713
75° 58′	.250	$.734\sqrt{a}$	$1.533\sqrt{a}$	$1.161\sqrt{a}$	2.692
71° 34′	.333	$.752\sqrt{a}$	$1.580\sqrt{a}$	$1.079\sqrt{a}$	2.656
63° 26′	.500	$.759\sqrt{a}$	$1.697\sqrt{a}$	$.938\sqrt{a}$	2.635
60° 00′	.577	$.760\sqrt{a}$	$1.755\sqrt{a}$	$.877\sqrt{a}$	2.632
56° 19′	.667	$.759\sqrt{a}$	$1.824\sqrt{a}$	$.812\sqrt{a}$	2.635
53° 8′	.750	$.757\sqrt{a}$	$1.892\sqrt{a}$	$.753\sqrt{a}$	2.645
51° 20′	.800	$.753\sqrt{a}$	$1.960\sqrt{a}$	$.724\sqrt{a}$	2.654
45° 00′	1.000	$.740\sqrt{a}$	$2.092\sqrt{a}$	$.613\sqrt{a}$	2.704
40° 00′	1.192	$.722\sqrt{a}$	$2.246\sqrt{a}$	$.525\sqrt{a}$	2.771
36° 52′	1.333	$.707\sqrt{a}$	$2.557\sqrt{a}$	$.471\sqrt{a}$	2.828
35° 00′	1.402	$.697\sqrt{a}$	$2.430\sqrt{a}$	$.439\sqrt{a}$	2.870
33° 41′	1.500	$.689\sqrt{a}$	$2.465\sqrt{a}$	$.418\sqrt{a}$	2.989
30° 00′	1.732	$.664\sqrt{a}$	$2.656\sqrt{a}$	$.356\sqrt{a}$	3.012
26° 34′	2.000	$.636\sqrt{a}$	$2.844\sqrt{a}$	$.300\sqrt{a}$	3.144
21° 48′	2.500	$.589\sqrt{a}$	$3.170\sqrt{a}$	$.288\sqrt{a}$	3.397
18° 26′	3.000	$.548\sqrt{a}$	$4.001\sqrt{a}$	$.119\sqrt{a}$	4.121

Example. — What are the best dimensions to give a ditch when the angle of slope is 45°, the discharge 36 cubic feet per second, and the velocity 4 feet per second.

Solution. — From formula

$$v = \frac{q}{a}, \quad a = \frac{q}{v}$$

hence, $a = \frac{36}{4} = 9$ square feet area.

Then vertical depth $= .740 \sqrt{a} = .740 \times 3 = 2.22$ feet.
Width at top $= 2.092 \sqrt{a} = 2.092 \times 3 = 6.276$ feet.
Width at bottom $= .613 \sqrt{a} = .613 \times 3 = 1.839$ feet.
Length of $ae = \dfrac{ab - cd}{2} = \dfrac{6.276 - 1.839}{2} = 2.218.$
Length of side $ac = \sqrt{ae^2 + ce^2} = \sqrt{2.22^2 + 2.22^2}$
     $= 3.14$ feet.
Perimeter $= 3.14 + 1.839 + 3.14 = 8.12$ feet.

Example. — What will be the best dimensions for a ditch when the angle of slope is 60°, the discharge 50 cubic feet per second, and the velocity 2 feet per second.

Solution.— $a = \dfrac{q}{v} = \dfrac{50}{2} = 25$ square feet.

The vertical depth $= .760 \sqrt{25} = 3.80$ feet.
Width at top $= 1.775 \sqrt{25} = 8.875$ feet.
Width at bottom $= .877 \sqrt{25} = 4.385$ feet.
Length of $ac = \dfrac{ab - cd}{2} = \dfrac{8.875 - 4.385}{2} = 2.24.$
Length of side $ac = \sqrt{3.80^2 + 2.24^2} = 4.41$ feet.
Perimeter $= 4.41 + 4.41 + 4.385 = 13.21$ feet.

Flow of Water in Ditches. — An approximate formula that may be used for ditches with earthen banks in good condition is the following:

$$v = \sqrt{\frac{100,000 \; r^2 s}{9\,r + 35}},$$

in which v = mean velocity in feet per second;

r = hydraulic radius = $\dfrac{a}{p}$;

s = the slope = $\dfrac{h}{l}$.

Example. — What will be the mean velocity of flow in trapezoidal ditch having a fall of 5.25 feet per mile, and the following dimensions: Top 28 feet; bottom, 10 feet; length of sides, 10.3 feet; depth 5 feet.

Solution. — Here, $s = \dfrac{5.25}{5,280} = .00099 +$, which call .001. Also, $r = \dfrac{95}{30.6} = 3.105$.

Then

$$v = \sqrt{\frac{100,000 \times 9.64 \times .001}{27.95 + 35}} = 3.91 \text{ feet per second.}$$
Ans.

In the example, the velocity is nearly 4 feet per second, would this be too great for the earthen banks of a ditch to resist without washing? The answer to this question can only be given by referring to the results of experience. It has been found that light and sandy soils cannot withstand a mean velocity greater than 2 feet per second,

while at the same time this velocity is sufficient to prevent plant growth and remove silt. In firmer soil,
velocities of 3 to 4 feet per second are permissible, but,
except in hard pan or very resisting material, 5 feet
seems to be the limiting velocity of earthen ditches.

In a district where it is proposed to build such ditches
there will be some examples of ditching, upon a greater
or less scale, by observing which an approximate idea
may be formed of the proper grade and side slopes to
be given to the proposed ditch, or if there are no ditches
the engineer must be employed.

Side Slope for Ditches. — The following table is given
by some authorities as the proper velocity for the maximum flow in ditches, and the proper slope to give the
sides of ditches in different materials.

Material of Sides.	Angle of Sides Degrees.	Slope = Base / Perpendicular	Maximum Velocity in Feet Per Second.
Hard rock	90	Vertical	10′ per sec.
Bedded rock.	90	Vertical	6′ per sec.
Schists	· 90	Vertical	4.5′ per sec.
Puddled clay	45	1 in. 1	4′ per sec.
Gravel 1.5 in. diameter. . .	40	1 in. 1.19	3.25′ per sec.
Gravel 1 in. diameter . . .	40	1 in. 1.19	2.25′ per sec.
Gravel, coarse sand. . . .	35	1 in. 1.42	.75′ per sec.
Gravel, fine sand	30	1 in. 1.73	. 5′ per sec.
Clay and soil	20	1 in. 2.74	.25′ per sec.

Besides the velocity and side slope there are other
factors that influence the choice of form of the cross
section of a ditch, and a certain depth will generally be

found more convenient or desirable than another. The following example will make this plain.

It is desired to establish the proper cross section and grade of a ditch under the following circumstances: The quantity of water to be conveyed is 150 cubic feet per second. A velocity of 2 feet per second is desired. The side slopes are to have an inclination of 1 vertical to $1\frac{1}{2}$ horizontal, and a depth of 3 feet of water is desired in the ditch. What should be the form and area of the cross section, and what the grade of the ditch?

Since the velocity is to be 2 feet per second, and the discharge 150 cubic feet per second, from the formula $a = \dfrac{q}{v} =$ the area will be $\dfrac{150}{2} = 75$ square feet.

To determine the form which this area must have it is necessary to know the bottom width of the ditch.

Since the vertical height is 3 feet, and the slope is to have an inclination of 1 to $1\frac{1}{2}$; ae, in Fig. 38, is 4.5 feet. From this data, let $x =$ bottom and $xe \times 2a$ or 9 = top, then the area $= \dfrac{3(x+x+9)}{2} = 75$ or $12x + 54 = 150$, and $x = 8$. ac will then $= \sqrt{3^2 + 4.5^2} = 5.4$; and $p = 5.4 + 5.4 + 8 = 18.8$.
Since

$$r = \frac{a}{p} = \frac{75}{18.8} = 4.$$

All the necessary data is now at command except s, and to obtain this insert all the data in formula

$$v = \sqrt{\frac{100,000\ r^2 s}{9r + 35}} = 2 = \sqrt{\frac{100,000 \times 16 \times s}{9 \times 4 + 35}}$$

$$4 = \frac{100,000 \times 16 \times s}{71};$$

hence

$$s = \frac{284}{1,600,000} = .00017$$

as the grade and this is equal to .00017 + 5280 = .897 feet per mile.

Depth and Velocity of Flow. — The depth of a ditch exercises considerable influence on the velocity of flow. For instance in the example, if the depth had been 6 feet, other data remaining the same, the bottom width would be $24 x + 108 = 150$ and $x = 1.75$. The sides would be 10.8 feet, and the wet perimeter 21.91 feet. The hydraulic radius 3.42, the square of which is 11.69. Then

$$2 = \sqrt{\frac{100,000 \times 11.69 \times s}{9 \times 11.69 + 35}}$$

and

$$s = \frac{4}{8350} = .00048.$$

This represents a grade of 2.53 foot to the mile, where with the previous depth it was .897 foot per mile.

On comparison these examples show that with a given grade and area of cross section, the velocity becomes greater as the depth increases, because, within certain limits, the hydraulic radius increases with the increase in depth. The limit is reached when the width of the ditch is equal to twice its depth. This condition is most perfectly fulfilled in the case of a ditch having a semi-circular cross section.

As a further illustration to show the effect of depth on velocity, take the semi-hexagon form, although it is an unpractical form for earthen ditches.

The semi-hexagon, Fig. 39, is inscribed in a semi-circle. Since the side of a regular hexagon is equal to the radius of the circumscribing circle, the relation

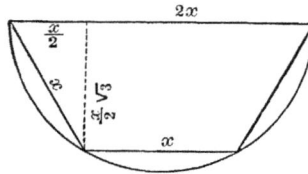

FIG. 39.

between the various parts shown in the figure exists. The side x, is required first, and the depth, which is $\dfrac{x \sqrt{3}}{2}$. Since the area in the first illustrative example is

75 square feet, $\dfrac{3 x}{2} \times \dfrac{x \sqrt{3}}{2} = 75;\ \sqrt{3 x^2} = 100$, and $x = 18.24$. In a ditch having the form of a semi-hexagon the hydraulic radius is equal to $\dfrac{d \sqrt{3}}{8}$ in which formula, $d =$ the diameter of the circumscribing circle or $2\ x$. In the example under discussion, therefore,

$$r = \frac{18.24 \times 2 \times 1.73}{8} = 7.9,$$

and

$$2 = \frac{\sqrt{100,000 \times 7.9 \times 7.9 \times s}}{9 \times 7.9 + 35}$$

or

$$s = 0.0067 \text{ or } 3.53 \text{ foot per mile grade.}$$

Siphons. — There are occasions when pipes are placed in ditch lines to carry water across valleys that can not be spanned by trestles.

Pipes used for this purpose become inverted siphons, but have received the name of siphons. Water will seek its level, but owing to the friction in the siphon and the necessity of having a head to create a flow, the entrance to the siphon must be higher than its discharge. Near Discovery, in Atlin, British Columbia, the water from a ditch is carried across the valley in a steel siphon, where it is delivered to a second ditch.

At Junction City, Trinity County, California, there has been laid 5700 feet of siphon pipe to carry the water over 280 feet depth of cañon. This was made in two sections: 2300 feet is No. 10 iron, 30 inches in diameter, and 3400 feet is No. 7 iron, 36 inches in diameter.[1]

At Cherokee, Butte County, California, an inverted siphon 30 inches in diameter is used to cross a valley 873 feet deep. This siphon pipe was nearly $2\frac{1}{2}$ miles long, and was the first large siphon used in hydraulic mining.

Siphons must be provided with expansion joints; be firmly anchored on the inclines ; have suitable air valves ; and laid with great care. When suitable care is observed in jointing and laying sheet-metal pipes, they will be as serviceable for siphons as in other hydraulic pipe-line situations.

[1] *Mining and Scientific Press*, Nov. 27, 1897.

CHAPTER VII.

GIANTS AND HYDRAULIC ELEVATORS.

Giants. — The nozzle from which a stream of water is projected in hydraulicking is called a giant.

With the introduction of stovepipe in the first stages of hydraulic mining it was found necessary to use a short piece of canvas hose in order to fasten the nozzle to the discharge end of the pipe. This was succeeded, as pressure was increased, by the gooseneck, a flexible iron joint formed by two elbows working over each other.

The improvement on this arrangement was the radius plate.

The Craig Monitor followed, and then the Fisher Knuckle Joint. Next came Hoskin's Dictator, and Hoskin's Little Giant, which at least has given the nozzles a name, as they are now termed "giants."

The Joshua Hendy Company, San Francisco, make what they term a double ball-bearing giant, while Hoskin's New Hydraulic Giant has various improvements over former styles. These improvements have been gradual, the more recent having increased their efficiency and convenience. Figure 40 represents Hoskins' New Hydraulic Giant.

The lever shown on the end is for moving the deflector, which throws the stream to any desired angle and moves the body of the giant. Horizontal and vertical motions are made with one joint, and this joint is protected so as to be durable. The nozzle butt is attached to the pipe so as to counteract the upward movement when working under great pressure. The nozzle is balanced by matching the notch in its flange with a corresponding one in the flange of the elbow. Where there is a downward tendency of the pipe, owing to low water pressure or small nozzle, use is made of the balancing attachment shown.

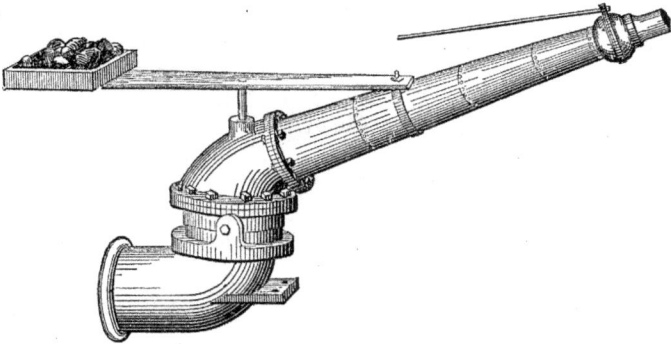

FIG. 40.

Practice has demonstrated that one giant with a large nozzle is better than several with smaller nozzles. The large nozzle proportioned to the pressure will do more work, and offers the economic advantages of concentrating the work, thus lessening the expenses.

The nozzles are constructed from 4 to 9 inches inside diameter, the inlets varying from 7 to 15 inches diameter to correspond.

To find the **spouting velocity** from a nozzle of any diameter, under any head or column or water pressure, and the amount of water which will flow per second through the orifice, proceed as follows: —

1. Find the area of the nozzle in square feet.
Area in square feet = diameter in inches × diameter in inches × 0.7854 ÷ 144.

Example. — What is the area in square feet of 1¼-inch-diameter nozzle?

$$1.25 \times 1.25 \times 0.7854 \div 144 = 0.0085.$$

2. Find the theoretical velocity, and multiply it by 0.80, the coefficient of friction caused by the rushing of water through the nozzle.

Theoretical velocity = $\sqrt{\text{Head in feet}}$ × 8.03.

Actual velocity   = Theoretical velocity × 0.80.

Example. — With a head of 25 feet, what is the theoretical and what the actual velocity with which water will spurt from an orifice?

Theo. vel. = $\sqrt{25}$ × 8.03 = 5 × 8.03 = 40.15.
Actual vel. = 40.15 × 0.8 = 32.12 feet per second.
Application to rule.

Question. — What amount of water will flow through a 1¼-inch-diameter nozzle under a head of 25 feet?

Rule. — Area in square feet of nozzle, × actual velocity in feet per second:

0.0085 × 32.12 = 0.273 cubic feet per second × 7.5

= 2.0475 gallons per second.

0.273 × 60 = 16.38 cubic feet per minute × 7.5
= 122.85 gallons per minute.

There are two general types of giants in use, namely

the single- and the double-jointed. The single-jointed nozzle is probably more efficient in manipulation, as it can be lubricated without turning the water out of the machine. The double-jointed nozzle has the advantage as far as the efficiency of the stream is concerned; it is also safer to handle under high heads. The double-jointed nozzle without a king bolt has a clear waterway,

Pipeman.

but on account of damage to its ball bearings at times, the king-bolt type is the better. The giant should be set securely on heavy timbers and bolted to timbers and the timbers to rock if possible. In any case stakes should be driven where there is gravel instead of rock, either through the bed piece or against the bed piece.

The nozzle is liable to "buck" if not securely anchored and damage both the pipe man and the attachments.

The giant must be lubricated so that it may be turned readily by the deflector, either right or left or up or down. The tendency in recent practice is to make a long smooth nozzle, and thus prevent the spraying as much as possible. In order to obtain the best results from any nozzle, air must be kept out of the pipe line, for which reason the pressure box, the air valve, and the standpipe are adopted, as already explained.

Hydraulic Elevators. — The Evan's gravel elevator, which is herewith illustrated through the kindness of the Risdon Iron Works Company, is used for disposing of tailing where sufficient fall for tail sluices is not available, or where it is impossible to run bed rock sluices. Hydraulic mining, now being under United States supervision, has, as before mentioned, received new impetus in California, where a demand for suitable machinery necessary to overcome obstacles is increasing; consequently, this machine, like the improved dredgers, comes very opportunely.

The principle of the elevator is that of a steam injector, or where the velocity of the water flowing up through an orifice is sufficient to cause a vacuum and hence a suction through a tail pipe. It is, of course, necessary to have a higher head[1] for such machines than is merely necessary to lift the water to a certain height, for friction of the water and the weight and friction of the gravel, together with gravity acting upon the whole mass of water and gravel, must be overcome.

Besides the motive-power pipe, D, Fig. 41, there are four other openings in this elevator, A, B, C, E. B

[1] About five times the head, over the lift.

Showing Evans' Elevator.   I.

and *C* are termed auxiliary suctions, which allow the
water and material to enter at the back of the seat, thus
reducing the wear and tear on the machine. The
auxiliary suctions can have their tail pipes extended to
any distance beyond the elevator proper, as shown in
the page cut, and thus be used for draining bed rock,
below the sluice connecting with the main elevator
opening, *A*. This may be very advantageous at times,
and can be carried on without interfering with the main
work of sluicing.

The auxiliary openings also increase the efficiency of
the elevator, by allowing the proper amount of air
and material to enter when the
main suction, *A*, becomes choked
or for some other cause is unable
to do its duty.

This feature economizes water,
which must otherwise be turned
off or run to waste while the
obstruction is being removed or the
difficulty obviated.

"Evans' elevators in New Zeal-
and, with less than 400 miner's
inches of water (600 cubic feet),
under a head of 225 feet, lifted

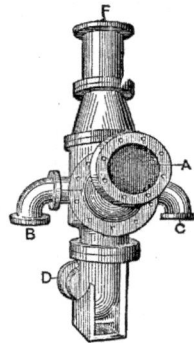

FIG. 41.

sand and gravel to a height of 52 feet at the rate of 2400
tons in 24 hours."

"Other similar work was carried on for years, elevating one acre of ground per month, varying in depth from
30 to 35 feet."[1]

[1] Mr. R. S. Moore.

The elevator to accomplish this work used 250 miner's inches, and raised its own water, the water coming from the giant at the rate of 1687.5 gallons per minute, and the material the giant washed out to it.

The expenditure of 223 H.P. to accomplish the work of 74.5 H.P., thus obtaining but an efficiency of $33\frac{1}{2}$ per cent of the power expended, does not at first glance seem economical, but when it is considered that 47 per cent of the power is used by the water in raising its own weight and that $19\frac{1}{2}$ per cent is employed in overcoming friction the machine as a pump becomes satisfactory.

Elevators are connected so as to be permanent, they may be arranged, however, so as to do their own sinking to bed rock, a commendable feature, when it avoids the necessity of making a pump by hand, which may require timbering and pumping arrangements, as well as a diver, to connect the elevator.

The excavation necessary in placing a 16-inch elevator at the Golden Feather River Mine, Oroville, California, was 4 square feet, while the previous year an old style elevator required 128 square feet of excavation and the services of a diver to place it in position.

The Evan's elevator was fitted up, lowered to the bottom of the river, and set at work in twelve hours' time.

The machines could be proportioned to elevate all the gravel which one giant could wash and sluice, were the material of a proper size to go through the throat of the machine. The area of the throat will depend upon the water available, and should be proportioned to the

Showing Evans' Elevator. II.

average size of the stones in the gravel bank. The latter is an indeterminable quantity, consequently screen bars or "grizzlies" are placed in the sluice to allow only certain-sized material to pass through into the sluice going to the elevator. Where water is available or lift slight the throat may be proportioned to accommodate large stones; the largest throat on record is for stones which will pass a 9-inch screen.

This is a subject of much significance, for the object of such machines must be in part, if not wholly so, to raise the greatest amount of material possible from the workings and put it out of the way once and for all.

Mining men thoroughly understand the importance of the preceding clause, and it has been stated to the writer that since the introduction of the Evans' elevator mining men are now seeking propositions which require an elevator, although heretofore, owing to heavy cost, weight, and inefficiency of old-style elevators, they would not consider them.

Mines, which, with former crude machinery, were unable to pay expenses, have by the use of these new machines been turned into dividend payers.

There are two instances where these machines have been able to elevate with a $2\frac{3}{4}$-inch jet from the motive-power pipe all the sand, gravel, and water it was possible to bring down the sluice on a 2 per cent grade. The giant had a $2\frac{1}{2}$-inch nozzle and used 1687.5 gallons of water per minute, while the elevator used 3187.5 gallons of water per minute, and raised water and material 52 feet. The highest elevation on record is 70 feet.

The Golden Feather Company, who are very large river operators, dammed the Feather River at Oroville with head and foot dams 1½ miles apart, the object being to work the gravel in the bottom of the river. The river at this place is between two and three hundred feet wide and from twenty to thirty feet deep. In order, therefore, to reach this gravel bed, wing dams were made and the water course changed, and finally the water between the dams pumped out. To effect the latter, two Evans' elevators were set at work and accomplished the pumping in 2½ days.[1]

The data required for estimating the duty for such elevators, and which the makers require are:

1. Quantity of water available. — This must include the amount of water the giant will use, and the remainder will only be available for the elevator.

2. The head of water in feet. — By doubling the size of a nozzle under a given head of water there is 4 times the quantity of water passed in a given time, while with 4 times the head but twice the quantity is passed by the same nozzle.

3. The distance the elevator must lift. — Usually but ⅕ the head can be counted upon, the head being used up in overcoming friction and the power which the weight of the column of water and material of the suction pipes have in retarding its flow, together with its own weight and friction.

4. The distance from bed rock to the top of bank. — If placed on the bank the elevator must raise a column

[1] Risdon Iron Company.

of water and material equal to the distance between
the throat and the level of the water. On the other
hand, if on bed rock, the weight of water and material will assist the elevator.

5. The largest size of gravel to be elevated.

There are three hydraulic elevators in the market the
Hendy, the Evans, and the Campbell, the latter is said
to be adapted to higher pressures than the others, and
has been successfully used in British Columbia and
Alaska.

Where it is possible to obtain water under pressure
low-lying placers have been readily worked by the aid
of elevators.   To attain good results, however, with such
apparatus the minimum amount of water should elevate
the maximum amount of gravel, and this can only be
attained by properly proportioning the elevator to its
pressure and lift.   The velocity of the water in the
pipes should not exceed 5 feet per second.   At Breck-
inridge, Colorado, there are two elevators working gravel
deposits with 160 feet head and lifting the material 42
feet above bed rock.   This is excellent work, as a 20-
foot lift to every 100 feet head of water is the maker's
rule, and in hydraulic practice the engineer considers a
lift of 15 feet for every 100 feet head sufficient for calculations.   Under favorable conditions the cost of working elevators is 5 cents per cubic yard, which includes
the hydraulicking.   A good elevator can handle from
1000 to 2000 cubic yards of dirt in 24 hours, provided
it is supplied in a reasonably uniform manner.   There
is another elevator, termed the Ludlum, which consists
of a giant nozzle fastened in a pipe in such a way as to

form a vacuum in the throat of the apparatus.   It does not differ materially in details from those mentioned, except it is not as scientifically constructed and is simply a round pipe affair.   It can do fairly good work and has been employed to some extent.

CHAPTER VIII.

EXPLOITING PLACERS.

Placer Mining Investments. — People have invested in placer mines under the impression that all that was necessary was to dig the dirt, and the water would save the gold. They were not aware that assays that included black sands were not to be relied on, as the gold in black sands can not be saved by hydraulicking. They were not aware that fine gold could not be saved with any reasonable degree of surety, nor did they take into consideration the fact that there are tricks in all trades. It never seems to occur to such individuals that mining is a science, and that to be proficient in that science one must have both the experience and the the theory of the subject at command. If a person purchases real estate, he hires a lawyer to examine the titles; if he wants his teeth fixed he goes to the dentist; if he is sick he calls on a physician; but when he goes into mining he takes the word of any plausible talker; and invests his money on the strength of a report made by a stranger. He assumes that mining is a gamble and that the successful mine investor struck luck. This is not the case, and he will find on enquiry that the successful mine owner, hires a mining engineer who has a reputation to sustain as a careful and reliable man in his profession.

There have been many failures in placer-mining

ventures, some of which were due to the lack of gold, or to the gold being so fine it could not be saved, while other failures were due to the incompetency of the management. A mining engineer, versed in the details that enter into the successful operation of placer-mining, has never, to the writers knowledge, failed in this business, for which reason a mining engineer of some reputation should be retained as consulting engineer until the business is on a paying basis.

Water Required for Hydraulicking. — The quantity of water required for washing dirt must be determined in a measure by the kind of gold in the dirt. Clear, sandy gravel will not require as much water as clay or cement gravel, and it is possible to use too much water in clay washing.

The size of a sluice depends upon the grade and kind of the gravel, also upon the water used and its duty. The duty varies. From the large amount of data received and tabulated by Mr. Bowie there is nothing absolute which can be placed as a rule, therefore, close observation must determine the quantity. According to Mr. Hall, of California, 3.6 cubic yards of dirt were moved by 1.5 cubic feet of water, for 24 hours' duration; this is equivalent to 2160 cubic feet of water to move 97 cubic feet of gravel, or 22 cubic feet of water to move 1 cubic foot of material. A. J. Bowie has tabulated 18 cubic feet of water to move 1 cubic foot of gravel at North Bloomfield, and 56 cubic feet of water to move 1 cubic foot of gravel at La Grange mines. In the former instance the grade was 8 per cent and the gravel light; in the latter place the grade was 2 per cent and the gravel the run of the bank. At North Bloomfield the sluices were 6 feet wide by 32 inches deep; at La Grange they were 4 feet wide and 30 inches deep. The height of the banks varied from 100 to 265 feet at North Bloomfield, and from 10 to 80 feet at La Grange, a difference that would exert considerable influence upon the quantity of material the water could come in contact with, and, therefore, mine.

The Cost of Hydraulicking. — The yield of the gravel at North Bloomfield was 7.75 cents per cubic yard, the cost of mining 4.1 cents per cubic yard. The yield per cubic yard of gravel at La Grange was 10.19 cents, the cost of mining 6 cents per cubic yard. The cost of mining at these two mines would analyze about as follows:

Labor	60 per cent
Supplies	17 per cent
Water	13 per cent
Office	10 per cent

Ground carrying 3.99 cents per cubic yard has been worked at a profit at North Bloomfield. With such a small margin to work on it is evident that skill and executive ability must be provided from the pipeman up.

In Idaho a placer having less value than 2 cents per cubic yard has been worked at a profit when others much richer have proved failures.

In the Seward Peninsula, where the gold is coarse and the water cold, no mercury is used in the sluice-boxes, hence all fine gold passing through the short sluice is wasted. That mercury can be used to advantage under these adverse circumstances has been proven. In most placer-mines where sluicing is practiced mercury is considered to be one of the most potent factors in collecting the gold. It must be used with judgment, however, otherwise it will be lost, and amalgam as well. Under the most favorable circumstances there will be a slight loss of mercury, which will vary according to the length of the sluice, and the care used in charging the riffles.

Charging the Sluices. — Before mercury is placed in the sluice riffles, dirt and water is allowed to run through the boxes for several hours. Too much water should not be run since the object is to pack the sluice and prevent leakage, and this can be best accomplished by the water and material moving slowly in the sluices. After packing has been accomplished, the water is turned off, mercury is poured into the riffles at the head of the sluice, and afterwards in all the riffles and undercurrents on the line. The head riffles are charged with more mercury than those further down the line since they will have more gold in contact with the mercury. The quantity of mercury to use in a sluice depends upon the quantity and kind of gold to be saved, and upon the length of the sluice. An 1800-foot sluice was charged with 900 pounds of mercury, while 150 pounds or two flasks are considered sufficient for a 240-foot sluice. Mercury should not be strained through a cloth, or spattered in pouring into the riffles, but should be poured from a cow's horn, the small end of which is sawed off so as to be opened and closed with the finger. If care is not taken in charging, little globules of mercury will be formed, and these will float away.

There are times when mercury becomes sluggish in its action owing to the impurities in the dirt, or to the quantity of gold it has amalgamated. Such mercury needs quickening by the addition of fresh mercury in small quantities, and this is accomplished by commencing at the top of the sluice and adding about 4 pounds to the first 50 feet; the next day adding 3.5 pounds to the next 50 feet, and so on until the end is reached, when the head of the sluice is again charged. Some charge every alternate 50 feet daily, but this is a matter that must be decided by the amalgamator who examines the mercury in the riffles with his fingers, and from its condition of pastiness judges its need of quickening.

Before mercury is placed in a sluice box it should be cleaned, and the best method of doing this is by retorting. Some use sodium or cyanide of potassium for cleaning; however, these chemicals are not so effective as the retort. If the mercury becomes foul, a clean-up should be made as soon as possible and the mercury retorted. The condition of the mercury can be ascertained by panning the sand from either a baffle board riffle or from the end of the sluice. When panning, a clean iron pan should be used, or one that has been thoroughly burned out to remove any mercury that may have been left in it from a previous panning. These remarks are more applicable to a small than a large operation, since the latter may not be cleaned up for a whole season, consequently the loss of amalgam and quicksilver will be considerable, even although fresh quicksilver is added daily. On the supposition that if mercury is freed at the top of the sluice it will be caught later on, long sluices are always economical; in fact, there was less loss of mercury in an 1800-foot sluice than in a 360-foot sluice; however, the two sluices were not carrying the same kind of dirt. Where sluices are too short, and the grades too high, or the sluices poorly constructed, or the ground washed contains much clay, talc, or other impurities, the loss of gold and mercury may amount to from 10 to 40 per cent.

If there are undercurrents in the sluice line they are charged at the same time the sluice boxes are given mercury, the quantity depending upon the width and length of the undercurrent. A good saving of gold by sluicing depends upon the skill of the operator, and may be placed at from 70 per cent up. This has reference to free gold where conditions are favorable for sluicing.

The Clean-up. — After the formation of amalgam, which is brittle compared with mercury, according to the amount of material it has absorbed, there is danger of loss by its floating away, and this means a loss of mercury and gold.

To avoid this, "clean-ups," or a collection of the mercury, gold, and amalgam, should take place as frequently as possible.

The cleaning-up process may take place in sections or the entire length of the sluice, commencing by washing out the bed-rock tunnel or the ground sluice with water, taking up their pavements, then washing the blocks and floor of the sluice boxes down to the first riffle. At this latter point all amalgam and gold washed down is taken up with an iron scoop and placed in an iron pail. The next section of sluice floor is now removed and treated the same way, until that portion of the sluice to be cleaned up has been gone over. The little water which was used to wash the blocks is now turned off, and the cleaning of the cracks and nail holes, termed "crevicing," is commenced by using silver spoons, to which the mercury and amalgam cling. This process having been gone through, the blocks are put back into the sluice and gravel washing commenced once more. The time occupied in cleaning up will depend upon the number of men put at it. Within 200 feet of the head of the sluice probably three-fourths of all the gold will be found; but smaller quantities will be found nearly to the tail sluice, depending upon the nature of the dirt washed and the kind of gold. The tail sluices are cleaned up only at the end of the season. The amalgam collected is now retorted and the quicksilver distilled from the bullion is collected.

Certain quantities of quicksilver will be lost in the sluices and in distillation. In the first instance the loss will be directly proportional to the quantity of water used and material washed, the grades being the same. If clean-ups occur within reasonable periods the length of the sluice will not be a factor, but otherwise it must be, in catching fouled mercury or amalgam particles. There is no method of determining the actual loss of gold, because there is no way of arriving at the absolute amount of gold in the deposit; it is, however, a certain percentage of the gold content.

The bed rock is cleaned by allowing a stream of water

to run through the channel until it is clear. Muddy water is said to aid in cleaning gold from the bed rock; however that may be, clear water is preferred, and crevicing may be needed in seamy rock to get the gold from cracks. When the bed rock is clean and the water runs clear through the sluices, the large volume of water is turned off, and a 1-inch rubber hose without nozzle used to wash down the blocks as they are taken up. The water should not be under much pressure, and in a wide sluice, from 4 to 6 feet, a hose on each side would materially hasten matters. The blocks as they are washed are placed on the sides of the flume. The hose may now be used to wash down the material on the bottom of the flume, or the material may be loosened with hoes and shovels, and about 2 inches of water let in for the same purpose.

One row of blocks is left at convenient distances apart in the sluice in order to hold back the mercury, gold, and amalgam, and as much as possible is taken out at this place. These blocks are next removed and washed, after which operation the washing of the sluice floor is continued to the next section, and so on. When the operation is finished the clean-up gang dig the amalgam out of nail holes and crevices, and after this has been accomplished the blocks are replaced. In some cases there is a clean-up every 20 days, in others once in two months, and again in ground sluices but once a season.

Small operators will probably clean up every week. It is probable that all operators should clean up when their floor lining commences to wear in grooves, and if large operators cannot afford to do this on account of their water supply, they can at least clean up a part of their sluices daily and run at night.

Amalgam kettles made of ordinary sheet iron, or better yet, cast iron kettles porcelain lined, are used as receptacles for the clean-up. The quicksilver and amalgam are stirred in these kettles in order to separate the sand and other foreign substances which float to the surface of the mercury and are there skimmed off. (In a clean-up in North Carolina, nails, pennies, and quantities of bird shot were found.) The remainder of the residue contains considerable amalgam and some sand, and this is worked up in a pan or rocker until concentrated. The concentrate is then ground in iron mortars with some quicksilver, and strained, the dry amalgam being treated in iron retorts.

Retorting the Mercury. — Where an operation is small and the clean-up takes place frequently, a small retort with a pipe nozzle can be used advantageously. In such cases there may be no necessity for straining off the mercury, and the amalgam and mercury may be charged into the retort together. The retort should be well chalked inside, with a clay lute to the cover, and the nozzle of the pipe should be immersed in a pail of water. With this arrangement, if there is no condenser, the pipe should be wrapped with gunny sack, and a stream of water allowed to run over the sack into the pail. By watching the end of the pipe, it is possible to see when the last of the mercury has left the retort, and when this takes place the retort is lifted from the fire and gradually allowed to cool before the cover is removed. The gold will be found in the retort in an impure condition.

Where large quantities of amalga mare to be retorted bricked-in retorts similar to that shown in section, Fig. 42, are used. These are made of cast iron with a movable cover at the back end and connected by a pipe with a condenser. The heat from the furnace passes over and under the retort and makes its exit through a suitable flue. The flue must be the entire length of the retort so that the heat will be uniformly distributed.

Before putting the amalgam into the retort the latter

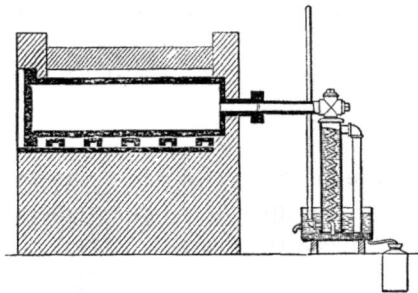

FIG. 42.

should be coated on the inside with a thin shee of clay to prevent the amalgam from adhering to the iron. The amalgam should then be carefully introduced and spread evenly. The pipe connecting the retort with the con-denser must be cleared of all obstructions and the amal-gam should be so spread that by no possible mischance can this pipe become choked, as an explosion would probably result and fill the retorting room with the poisonous fumes of mercury. To avoid danger the heating should be very slow at first. After the cover has been put on with a luting of clay and securely clamped, the fire is lighted and the heat is gradually raised to a dark red, that temperature being all that is necessary to volatilize the quick-silver. Toward the end of the operation, the heat is raised to a cherry red until distillation ceases. The retort is then allowed to cool, and, when cold, is opened. During the operation, the condensing coil is kept cool by a continuous supply of fresh water entering at the lower end of the cylinder that surrounds the coil, while the discharge of warm water takes place at the upper end. The retort bullion

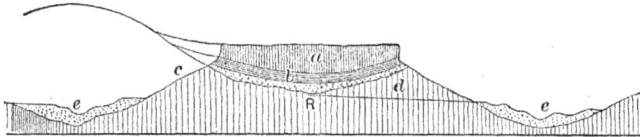

FIG. 43.

is cut or broken in to pieces and melted in an annealed black lead crucible and the gold cast into bars.

Drift Mining. — In California, placer deposits have been formed by ancient rivers, and these rivers were afterwards made the beds of lava streams that completely capped the placers as shown at a in Fig. 43. Where the deposits are deep and wide tunnels, d, are driven through the rim rock, c, to strike just below the center of the ancient river bed, b. The object of this tunnel is to drain the deposit and to transport the ground to the valley where it may be sluiced. Such tunnels may vary from one quarter of a mile to a mile in length, depending upon the width of the deposit and the thickness of the rim rock. Care must be exercised in giving them the proper grade for drainage. If the tunnel should be driven so as not to end at R, but to strike the deposit higher up, its object would be only partially successful.

At R another tunnel is driven in d under the river bed at right angles to d, and at distances of 120 feet apart risers, r, Fig. 44, are put up to the deposit for the purpose

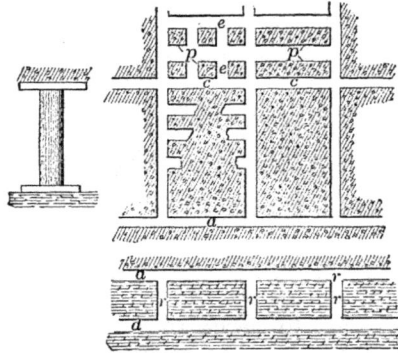

FIG. 44.

of drainage and to act as chutes for the mined ore. From these mill holes, levels are driven right and left in the deposit until the rim rock is reached. A second tunnel, a, Fig. 44, parallel to the one in bed rock, is driven in the deposit, in order that air may circulate in the different workings, and in addition to this, breakthroughs, b, between two parallel cross levels, c, are made at stated intervals as shown in the plan. Every 60 feet on the cross levels rooms are turned, and thus the number of working faces are increased. When the cross levels reach the rim rock the pillars, *p*, are driven through until they are cut up into small pillars, *e*. These cannot be recovered readily, and are left, unless they carry so much gold that it will pay to timber or to adopt either the caving method, as practiced in iron mining, or long wall mining as practiced in coal mining. In the rooms single posts mounted on plank sills and furnished with plank cap pieces are used for holding up the gravel. In the levels four-stick timbering with lagging is used to support the excavation, for if the levels close the work must be abandoned until new levels are driven.

It is, of course, understood that before drift mining is undertaken there must be prospecting done to ascertain the value of the deposit, and whether the returns will be more than the expenses incurred, to drive tunnels in rock, timber levels in gravel; mine, gravel and transport it to the sluice box at the mouth of the tunnel; make arrangements for water, construct ditches, flumes, and sluices, and finally the incidental charges arising at all mining operations, — are items that the gold from the mine must more than pay.

There have been a few cases where drifting in ancient river beds could be carried on without driving a tunnel. Such mines are located in places where the modern rivers have cut across the ancient rivers instead of running parallel with them.

Shafts and inclines on bed rock have been put down to work ancient rivers, but are not as cheaply worked as tunnels, since a hoisting plant is required to raise the material to the sluices at the surface, and further, pumps must be used to keep the mines drained.

In sinking to bed rock through such deposits every foot of ground won must be timbered, as the sides of the shaft will not stand unless supported. Often quicksand or running ground is encountered, and this being difficult to penetrate, it may necessitate the abandonment of the shaft. Inclined shafts are generally sunk so as to follow the rim rock, since in this position there is a firm foundation for the sills of the timber sets. Should an incline be put down in the gravel bed, the timbers would soon be moved out of alignment and the shaft ruined.

In the Atlin, B. C., district, shafts are sunk about 90 feet to bed rock, through material that contains no stone capping.

The bed rock has a slope of from 5 to 7 per cent. From the shaft drifts, approximately 100 feet wide and 450 feet long, are driven. The shaft is timbered from top to bottom and is operated by a steam hoister. The heaviest timbers procurable are used for posts and caps in the drifts and levels. Within a year 15,000 cubic yards of pay dirt were hoisted. The excavation made required 2600 sets of timber and over 25,000 lagging. The total cost of wood for timber and fuel at this mine was about $10,000, but as the gravel ran $4 to the cubic yard a snug profit was made, even against such odds as a creek running into the mine, owing to a cave and continuous pumping. The mine dirt at this place is loaded into cars, raised to the surface, and sent to the sluices in the same car over a short surface track.

This is but one illustration of drifting in that district, under a blue clay over which runs a creek.

In some cases where the placer deposit and the material above it were thick and conditions would permit, a tunnel has been driven through a side hill to act as a sluice and carry the débris to a suitable dumping ground. The valley into which the tunnel empties must be lower than the one containing the deposit. After a tunnel has been driven a shaft is sunk in the deposit to meet it. Down this shaft the surface dirt is washed and flushed through the tunnel. When bed rock is reached by washing away the dirt from about the shaft, a rock sluice is constructed from the bank to the tunnel and regular hydraulicking carried on. Sluice boxes and riffles are constructed in the tunnel whenever the bed rock is so soft as to wear rapidly. Bed-rock sluices may be constructed in hard rock, but since they cannot be cleaned up so readily or so often, wooden sluice boxes are preferred.

Blasting Gravel Banks. — Hard cemented gravel is not readily broken from a bank by water pressure, and as an aid to the water powder is used to shatter the ground. For this purpose it is necessary to drive a level in the bank and then cross levels at right angles to the ends of the level. The position of the level relative to the height of the bank is of some importance, as well as the stratum of dirt in which the level is driven. It is not advisable to drive on bed rock, as a certain part of the blow resulting from a blast must be delivered downwards, and on bed rock it performs no useful work. It is better to drive the tunnel in the cemented gravel than in a stratum above or below it, as the material is very difficult to shatter. The practice in California is to drive the level at two-thirds the height of the bank, for instance, a bank 120 feet high would have the level driven at a height of 80 feet from bed rock.

The object of blasting gravel banks is to shake the material, and loosen it for the giants to wash down. Nothing would be gained by using so much powder that the earth would be thrown out like rock in rock blasts, from the fact that the giants cannot wash broken-down material to the sluices as advantageously as when the material is in the bank.

The kind of explosive to be used in such blasting operations is a matter of convenience rather than choice, although in most cases black powder is the better. There is an exception to the above statement, namely, where water is the tamping, and then dynamite should be used.

Dynamite containing from 40 to 50 per cent nitroglycerine is preferable to higher grade explosives, as they furnish a rending rather than a shattering blow, approaching somewhat in execution the blow produced by black powder.

Experience must be the guide in the quantity of powder to use in bank blasting, for instance, some banks require 1 pound of powder for each 50 cubic yards of gravel, while other banks require 1 pound of powder for each 20 cubic feet of gravel. In a new bank it would be well, therefore, to take an average between these two extremes, or 1 pound of powder for every 35 cubic yards of dirt to be shaken. If this quantity can be reduced it should be, but if it does not accomplish its work properly, it must be raised.

The quantity of powder required for a blast is calcu-

lated according to the yardage affected, for example:
What charge of powder is required to shake a bank
50 feet high, the main drift to be 75 feet long and the
two cross drifts 30 feet long each way, and at right
angles to the main drift? The number of cubic yards
shaken will be about the dimensions furnished by the
solid defined by the data in the example:

Thus

$$\frac{75' \times (2 \times 30') \times 50}{27} = 8333 \text{ cubic yards,}$$

and $\frac{8333}{35} = 238$ pounds of powder as the charge.
The length of the main drift heading should be longer
than the bank is high, thus making the distance to the
surface the line of least resistance to the blast. The
charge should be placed about half and half in each end
of the cross drift and walled in. The tamping must be
done carefully and carried to the main drift, and at
this point a wall should be built and more tamping
added. In case the explosive gases are given space to
expand, they will not be as effective as when they make
their own space for expansion, for which reason ram-
ming the tamping is not objectionable. Wherever
possible the charge should be fired by an electric bat-
tery and not by fuse. In case the explosive is black
powder, then a piece of dynamite with a fulminate cap
inserted should be used as a detonator instead of relying
on the fuse to fire the powder.

Fuse is an uncertain quantity, for it is apt to be
cracked in unwinding and jamming; again, there is no
certainty of the charges at each end of the cross drift

being discharged at the same time; and further, it will be liable to injury during tamping operations. If it must be employed, then double or triple tape fuse is laid in duplicate lines in order to make sure of a successful blast. Small blasts are less economical than large ones; however, a series of small blasts are more effective than one large one. It will be policy, therefore, in case the face of the bank will permit, to drive two or three drifts about 75 feet apart and make the cross drifts at their ends 25 feet long each way. Each cross drift should be loaded the same and the entire series fired at once.

In some cases there is one drift, and from this two or three pairs of drifts are driven at right angles. The powder must be now proportioned to the yardage each charge is to break. The charges must be properly tamped and all fired at once. Judgment must be used in dealing with this kind of blasting if the maximum work is to be obtained with a minimum expenditure of powder. Experiment and close observation of the work accomplished will prove more satisfactory than any definite rules that may be laid down for the use of powder.

Mining in Alaska. — In Alaska the summer season is too short to thaw the ground to any great depth; further, in such cold climates the surface is usually covered with moss that prevents the sun's rays from penetrating the earth. When the placer gravel is not over 12 feet thick the moss and all other vegetation is stripped from surface and under such conditions the sun will thaw the ground from 1 to 2 feet per day.

The muck does not thaw as fast as the gravel. However, as fast as a layer of dirt thaws it is removed by ground sluicing, until the pay streak is reached. The pay streak is washed in sluice boxes.

Where the placer ground is more than 12 feet deep, shafts are sunk to bed rock. The top or muck soil is picked and shoveled, but when gravel is reached it is either thawed by wood fires or by steam points. A wood fire in a shaft will thaw from 8 inches to 1 foot per day, provided there is not too much frozen water in the gravel. When bed rock is reached, drifting is commenced, and either wood fires or steam used for thawing. A good wood fire will thaw from 12 to 18 inches into the face, and a fire 30 feet long doing this amount of thawing is good work for two men on a 4-foot pay streak.

Steam is now more generally used in Alaska than wood for thawing, although the cost of coal is $25 per ton, on the Seward Peninsula. The plant consists of a boiler, connected with steam pipes leading to the face. At this point a manifold is used, to which several steam hose are fastened. To the free end of each hose a nozzle is attached, called a "point." The point is from 5 to 6 feet long, about $1\frac{1}{4}$ inches in diameter, with jet holes $\frac{1}{16}$ inch in diameter at the pointed end, and a drive end at the other. The hose is attached to a tee near the drive head. The points are placed about 3 feet apart in a breast, and are gradually driven in as the steam thaws the ground. They are allowed to stand 5 or 6 hours under a steam pressure of 20 or 30 pounds per square inch, when they are disconnected. The ground is picked down, loaded into buckets, hoisted to the surface, and dumped on a pile as fast as it is thor-

oughly thawed. As soon as the spring thaws commence, the pile is sluiced in boxes. Scrapers attached
to horses are used to carry dirt from the pile to the
head of the sluice, and on account of the flat ground
scrapers are also used for spreading the tailings. As
a rule, very little timbering is required in frozen gravel;
in fact, it is safer than solid rock. It is customary to
drive rooms 400 feet long and from 30 to 70 feet
wide without any timber to support the roof. In the
tundra mines on Little Creek the gravel is frozen to bed
rock which is at a depth of 120 feet.

CHAPTER IX.

EXPLOITING PLACERS. — (*Continued.*)

Mining in North Carolina. — There is a belt of chloritic schists extending from Georgia to Nova Scotia in a northeasterly direction. The schists, which are metamorphosed and impregnated with quartz stringers, have been assigned by geologists to Cambrio-Silurian times. The quartz in these schists is highly mineralized in places, and carries gold, silver, copper, lead, arsenic, antimony, and possibly other metals. In some localities this belt of rocks has been altered, particularly in North Carolina; while in other localities it is evidently unaltered, particularly in New England. Owing to the physical condition of these rocks in North Carolina they are in several localities termed placers, and are worked as such; but in order to work them profitably, methods that are economical and efficient must be adopted. While the placers are considerably richer than the majority of western placers, nevertheless they are not as readily worked.

The gold is in a fine condition, but that is not the entire reason why it cannot be recovered by methods that have proved efficient in the West. Experience has convinced most miners that hydraulicking is not suited to these deposits, owing to the clay and the impurities which they carry and the fineness of the gold.

The placers which cover wide areas in North Carolina have been a source of aggravation to miners, who have seen others that carry very much less gold worked at a profit.

To the uninitiated it would seem that it was merely necessary to turn on the water in order to become wealthy.

At the Edith mine in Catawba County nearly every phase of hydraulicking known has been practiced with indifferent success. Owing to a scarcity of surface water it was necessary, in order to mine and sluice the ground, to sink shafts; construct a large retaining dam, to form a settling pond; and finally pump the water from the settling pond back to the mine. To sluice the dirt to the settling pond it was necessary to drive a tunnel and dig ditches. In addition to this work, considerable money was expended for boilers, pumps, and water pipes.

While there was not sufficient gold recovered by hydraulicking to pay expenses, still there was enough to entice the owner to renewed experiments with the hope of final success.

The author believes he was the first to suggest the use of the log washer for such deposits, basing his belief on its fitness from a knowledge of the good work accomplished in washing phosphate rock, iron and zinc ores. Mr. Overton was the first to put the idea into practice, and originated the washer shown in Fig. 45.

The washer is in two units, each having a single log, driven by belts from the same line shaft. Each log is 8 inches in diameter, constructed of steel pipe, reinforced by wood inside to add stiffness. The paddles are arranged spirally as on any log washer, but differ from those usually adopted, being flattened cast-iron blocks weighing 9 pounds each.

The logs are geared to run at 300 revolutions per minute, at which speed the blocks act as hammers and break the material. (At the Catawba mine the speed is but 90 revolutions per minute.) The blocks are not intended to break hard rock, but medium hard rock. The troughs in which the logs revolve are semicircular,

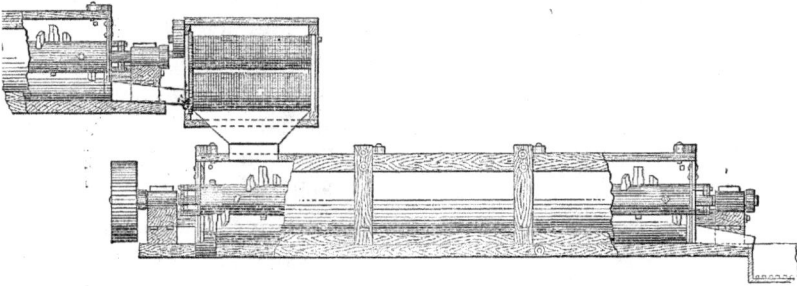

FIG. 45

made of ½-inch boiler plate and supplied with a wooden top that locks to the trough. The first washer is 18 feet long, the second 12 feet long, each being given a rise to the discharge end of about ½ inch per foot, the object being to form a larger receptacle at the entrance than at the discharge end of the trough. At the discharge end of each washer is a spout, *b*, leading to a rotary screen 2 × 4 feet; the first screen having ¾-inch openings, and the second one ⅛-inch openings. Below each screen is a trough. The first trough discharges the material that goes through the meshes of the first screen into the second washer, while the second trough discharges the material to the riffles.

The material to be washed is delivered through the hopper to the trough, where the coarser part is worked forward by the spiral arrangement of the paddles to the spout, leaving the gold and most of the heavy sand behind.

All material leaving the first box flows into the screen, which removes all stones larger than $\frac{3}{4}$-inch diameter. The remainder falls through into launder, and flows into the top of the second washer. The stones that did not go through the screen mesh, work out at the end and fall on a short endless belt that conveys them to a long endless belt moving in a direction parallel to the length of the washer. The long belt conveys the stones coming from both screens to the tailing sluice shown in Fig. 46. Each of the four washers at this mine is supplied with riffles 75 feet long and about 4 feet wide. The riffles are seen in the illustration to be on each side of the conveyer belt, and all discharge into a common tailing sluice.

The riffle floors are of 2-inch plank, with 2-inch diameter holes about 1 inch deep, bored in them. The holes which are staggered, are charged with a small quantity of mercury. Very little gold is caught in the riffles, the greater quantity being caught in the washers in about the following percentages:

First washer, 80 per cent of the gold.

Second washer, 18 per cent of the gold.

Riffles, 2 per cent of the gold.

Fig. 46.

FIG. 47.

182

Fig. 47 is an illustration of the Edith mine under the present working conditions, with a dirt bank 116 feet high. The log shanty in the excavation is the pump house of the main water supply, and is located over a shaft. From the bench of earth on which the pump house stands, drill-holes are put down, squibbed with dynamite, and blasted with black powder. The blast shakes the dirt so that when it reaches the loaders

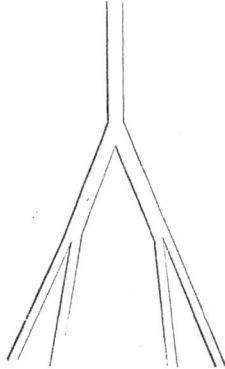

FIG. 48.

at the bottom of the cut it is in condition for easy shoveling.

Twelve men load 400 tons daily into cars. The dirt is hoisted up an incline and dumped automatically into a hopper, that is flushed by a stream of running water. The water carries the dirt to grizzles having 2½-inch spaces. All hard rock which is too large to pass through the bars is thrown out, while the smaller stuff is sluiced to the washers through the sluice box, constructed as shown in the plan, Fig. 48. There are 4 washers at this mine, each capable of washing 100 tons of dirt in 10 hours with 75 gallons of water per minute. It is, therefore, necessary to make as even a distribution of the material as possible, and this is accomplished as shown.

The value of the gold recovered is 60 cents per ton of dirt. The clean-up of the washers takes place once a week. It is performed by shutting down the mine for half a day and running clean water through the washers until it comes out clean at the riffles. The top is taken from the washers, and the material in them is shoveled into buckets and dumped in the clean-up box. There is a trough leading from the clean-up box supplied with Hungarian riffles containing mercury. The mercury in this trough seems to catch all the free gold, showing that the scrubbing and washing made it susceptible to amalgamation. Figure 49 shows the washing system before it was housed at another plant in North Carolina.

Steam Shovel Mining. — The problem presented to the Atlin Consolidated Mining Company was the excavating, hoisting, and washing of a 20-foot bank of compact yellow gravel. No natural dump existed for the storage of a large mass of tailings, and stacking and sluicing had to be resorted to. The plant shown in Fig. 50 consists of a steam shovel, a, having a dipper that has a capacity of $1\frac{3}{4}$ cubic yards; 18, 3-cubic-yard capacity side discharge ore cars b, having a 3-foot gauge, to run on a 30-pound rail track; an inclined plane, c, having a 30° slope, and terminating in a platform 45 feet above the bed rock. The cars dump their contents on the platform into chutes provided with grizzlies made of rails. The large material shears off the rails and falls to the stone dump; the small material passes through the bars into the sluice. The sluice is 4 feet each way and 144 feet long, set on a grade of 10 per cent. The first 48 feet are paved with 45-pound rails, placed longitudinally, the remainder of the

FIG. 49.

sluice with $3 \times 3 \times \frac{1}{2}$-inch angle iron forming cross riffles. The tail sluice and the extensions are block paved and given a grade.

The arrangement of the tail sluices is similar to that shown in Fig. 24, where there is a lack of dumping ground and Y's are constructed. The plant is worked by electricity, as far as hoisting and haulage are concerned. The shovel, however, is operated by steam, the digging engine having 100 horse-power, while the thrust and swinging engines are 30 horse-power each.

The plant successfully handles 1500 cubic yards of gravel per day, and employs from 30 to 40 men. The

FIG. 50.

water is taken from Pine Creek, which furnishes about 1500 miner's inches for washing purposes, while the electric power is purchased from a nearby power plant. No statistics are furnished in regard to the value of the gravel or the cost of working. However, the Guggenheims are interested in the company, which is a sufficient guarantee that it is not being operated at a loss.

Cableway with Self-filling Bucket. — In the Alder Gulch of historic fame, a placer deposit was once worked. The German Bar Mining Company, Virginia City,

Montana, believed it was worth reworking, provided they could excavate at low cost and transport the material to a tower of sufficient height to furnish a sluice and a

FIG. 51.

dump ground. The Lidgerwood Manufacturing Company furnished the plant for them, which consisted of power to run a radial traveling cableway, a pivoted tower and hopper, and a self-filling bucket.

The pivot tower, Fig. 51, had a large hopper, a, 40 feet above the ground, into which the bucket, b, discharged gravel to a 30-inch sluice, c, set on a 5 per cent grade. The sluice, which was 200 feet long, discharged its fine tailings 25 feet above bed rock, and its coarse tailings near the tower.

The pivot tower had a ball-bearing top, d, arranged to turn on its axis, and so allowed a second traveling head tower, a, Fig. 52, to move through an arc of 180°.

The latter carried the boiler, machinery, tower, and cable anchorage, and traveled on curved tracks. The excavation was made along radial lines, and thus a semicircular pit was worked out around the hopper tower. After each semicircular pit was excavated, the entire plant was moved forward and another pit made.

The Knight excavating bucket was developed to dig tough ground by shaving the top, and at the same time crowding right into the material. The bucket is supplied with teeth that strike the material as it is dropped by the fall rope. The illustration, Fig. 51, shows it, however, in the position it assumes when loaded.

The method of excavating is as follows: The carriage, with bucket hanging teeth downward, is run out on the cable, and the bucket dropped. The bucket strikes the ground teeth first, settles down on its bottom, and, as carriage continues toward the traveling tower with hoisting line slack, the bail falls into its natural position, the back catch automatically locking itself, ready for digging. When the carriage has reached a position much nearer

FIG. 52.

the tower than the bucket, the conveying-drum brake is thrown in, the carriage held stationary on the cable while the hoisting rope is tightened, thus giving to the bucket a long inclined draft, which enables it to fill. This draft may be varied by simply changing the position of the carriage with respect to the bucket.

The ease in changing the angle of draft is of the utmost importance in adapting the bucket to the material and depth of cut.

The bucket draws into the material, the strain increasing until the teeth are buried in the ground, when the leverman releases the conveying brake, allowing the carriage to gradually slip back to a position over the bucket, thus gradually changing the draft, the bucket meanwhile continuing to dig until the carriage is directly over it, by which time it is filled. It is then hoisted and conveyed to be automatically dumped into the hopper.

On level or difficult ground, the long draft, gradually decreasing, is absolutely necessary, and it can only be secured with the endless-rope system. Even in high banks the changing draft is of great value, as the bucket may be hoisted as soon as filled without having to drag it through all the material higher up the slope. The changes are made without stopping the engine, or motion of bucket, and effect a decided saving in time of filling.

The fine gravel dumped in the hopper, is washed through the grizzly, m, into the sluice, and thus separated from the large stones. The water comes to the hopper through a 12-inch diameter pipe, n, either from a side hill, ditch, or a pump. In this case it was a ditch that furnished a pressure of 40 pounds per square inch.

The bucket had a capacity of $1\frac{1}{2}$ cubic yards, and 400 buckets have been filled in 10 hours. As soon as the bucket made a channel to bed rock, it was used as a bed-rock sluice through which the top soil and fine material were washed. The water rushing through the cut towards the bucket carried the lighter material to a bed-rock flume having a grade of 1 per cent, which was not sufficient to move the gold-bearing gravel, and this was excavated by the bucket.

The grizzly was made of $\frac{1}{2} \times 3\frac{1}{2}$-inch iron bars with 2-inch spaces between them.

The boulders that pass over the grizzly stack up on either side of the tower, while the gravel that passes through the chute falls 5 feet to the sluice shown in Fig. 51. The sluice bottom is covered for 24 feet with 16-pound mine rails, the flanges being placed close together. The remainder of the sluice is covered with blocks 4×4 inches and separated crosswise by strips of wood 2 inches high to form the riffles. The labor force for this cableway consists of 5 men, — engineer, fireman, signal man, hopper man, and rigger. There are many placers in the South where cableways could be used to advantage to dig and transport material to log washers. There are other places in the West, particularly Nevada and Arizona, where they could also be used to advantage. In fact, the system of cableways has been developed until its flexibility adapts it to many kinds of placer mining.

CHAPTER X.

GOLD DREDGING.

Dredging has come into prominence within the last 15 years. New Zealand is the original home of the successful dredge, where it has been operated since 1886. On the Molyneux River, in New Zealand, there were at one time sixty dredges in operation, and the evolution of the present type was brought about by the experience originating in that country. The river bars gave indications of gold, and being at times rich, it was known that the river bottom must contain gold in paying quantities. The miners of the earlier days could work the shores of the river with spoons, which consisted of a bag laced or riveted around an iron frame and secured at the end of a long pole, so adjusted and weighted that it could be drawn along the bottom. When filled, or partly so, it was hauled up. Boats were next used with this spoon, and an auxiliary boat contained a rocker for separating the gold from the dirt. This dredging was the forerunner of the present bucket system of elevating.

"The first primitive vessel took the form of a couple of barrels surmounted by a timber platform, on which the dirt was shoveled by a man standing in the water, the dirt afterward being taken on shore and cradled." "The next dredge consisted of three canoes lashed together by a board platform and secured by ropes to

the shore to steady it. It was provided with the spoon
already mentioned for excavating. This contrivance was
the first pontoon dredge. The next step was to use
water-power to work the spoon, and where such power
was not available dredging was carried on by spoons
being raised by crab-winches worked by hand."

Mr. Ward, the inventor of the current spoon-dredge,
designed and worked successfully, in 1870, a bucket-
and-ladder dredge. The motive power for moving the
buckets he obtained from current wheels in the river.
This was followed by hand-power, then steam-power,
and electricity as practiced at the present time.

Dredging is one of the popular ways of recovering gold
where the depth of the alluvion does not exceed 60 feet
below water level or 20 feet above. The number of
dredges at work 10 years ago was about 60; at the present time there are said to be 500 at work in various
parts of the world. The ease with which placer ground
can be prospected, and the certainty of cost and recovery,
place dredging on a commercial basis almost if not
quite as secure as manufacturing.

Success or failure in any line of business depends upon
experience and watchfulness.

To assume that any one kind of dredge is suitable
for every kind of ground is a mistake, yet this is done,
and failures are recorded.

One kind of dredge is better adapted to one kind
of ground than another. For instance, where the gravel
bed contains few boulders and bed rock is soft, the
bucket dredge finds favor; or where the boulders are
large and the ground tough and cemented, the dipper

dredge is to be preferred; again, where the material is loose sand, as in some river beds, the suction pump is preferred.

To reduce dredging to a metallurgical proposition some system of sampling tailings to ascertain the loss that is occurring must be adopted. Samples for this purpose should take all the stream part of the time, and the sample so obtained should be concentrated in a rocker, and the gold extracted by mercury.

There seems to be a great diversity of opinion in regard to the proper method of saving gold on dredgers. Some operators prefer the sluice box, others prefer tables, and still others a combination of both. The latter method is probably the best, since no arrangements have yet been devised that will save all the gold; in fact, it is good work to save 70 per cent. Sluice advocates believe in long sluices, although J. P. Hutchins[1] states a case where material that passed through a 120-foot sluice was redredged and passed through a 30-foot sluice, with the result that the short sluice yielded as much as the long sluice, under adverse conditions. There has been very little improvement in the mechanism or construction of dredges, although the cost of working has been somewhat reduced by increasing the capacity. Where two or more dredges are working near each other a central power plant will considerably reduce the cost of the process, by decreasing the cost of handling fuel. No attempts have been made to save the black sands, that often contain values which run from one to ninety ounces per ton of sand, and that sometimes in addition

[1] Mineral Industry, 1905.

carry silver, platinum metals, and copper. Sands in the Caribou District, British Columbia, carried 64 ounces of platinum per ton of concentrates, worth $1920 at the present price $30 per ounce.

It is well known that heavy gold can be caught in sluice riffles, except when it is coated with an oxide of other material. Basing their position on this fact, many dredge operators do not use quicksilver in sluices, consequently lose fine and float gold. It is probably for the same reason that gold-saving tables have been discarded for sluices. Australian dredgers and Alaskan miners do not use quicksilver, although in the latter country it has been fully demonstrated that quicksilver will recover considerable gold from the tailings. If quicksilver is not needed, there is no fine gold in the placer, a condition of affairs that may be much doubted.

An objectionable feature common to all dredges, particularly bucket dredges, is that they cannot clean the bed rock thoroughly. Wherever an attempt is made to clean bed rock by hand, the water must be drained off subsequently to carrying the tailings beyond a point where they will not run back to the place recently excavated. It is probable that in many cases a suction pump could be used to advantage in cleaning bed rock after the bucket and dipper dredges have removed the greater part of the gravel.

The principal cause for the recent boom in dredging until it has become the rival of hydraulicking plants in California, is that the returns can be calculated with much certainty.

The difficulties in the way of successful recovery by

dredging are yet many, not in the commercial sense that dredging will not pay dividends, but in the metallurgical sense of saving a larger percentage of the gold. It has been found that a bucket dredge will recover 70 per cent of the values shown by the drill-holes, and that the recovery of from one to one and one-half grains of gold to the ton will pay expenses, the cost of operating ranging from 3 cents to 8.35 cents per cubic yard. The distribution of expenses in dredging operations are about as follows per cubic yard:

	High.	Low.	Cost Working 5 Foot Bucyrus Dredge Per Cubic Yard. Oroville.			
Pay roll . .	2.05	1.13	Lower. .	1.06 to 1.77	Average 1.415	cents
Power. . .	1.77	1.34	Repairs .	2.86 to 3.03	" 2.945	"
Repairs. .	3.80	.61	Labor . .	1.64 to 2.05	" 1.845	"
Sundries. .	0.73	.52	Expenses.	0.64 to 0.73	" 0.685	"
Taxes . .						
Total . . .	8.35	3.6c	Totals .	6.20 to 8.35	" 6.890	"

The cost of dredging depends upon the nature of the ground, and, other things being equal, the capacity of the dredge; since fixed charges practically remain the same whether 40,000 or 80,000 cubic yards of dirt are handled monthly. For instance, a $3\frac{1}{4}$-foot bucket dredge mined material for 4.892 cents per cubic yard, while a 5-foot bucket dredge mined the same material for 3.66 cents per cubic yard. In another case a 4-foot bucket dredge in a very hard compact cement gravel could not work for less than 8.7 cents per cubic yard.

The dredge is not a useful arrangement in a swift-flowing river, and is best suited to working river banks from the shore inland, in some cases to distances over a mile. Dredges working inland make their own floatway as a usual thing, although they are sometimes greatly assisted by giants or steam shovels that strip the top barren dirt above water level.

The construction of a dredge is of considerable importance. For instance, in a practically new country where no dredging has been done, exceptional care should be taken to investigate the kind of ground the dredge must handle, and the gold it is to save. Small dredges should be constructed in new districts, and then, if changes are necessary, the next dredge, which will be larger, can be equipped to meet the demands.

Dredges cost in proportion to their size and power. For instance, a dredge that will excavate and wash 40,000 cubic yards of dirt per month would cost approximately $40,000, while a dredge capable of handling 60,000 cubic yards of dirt per month would cost $60,000. The increased cost is due to the enlarged hull and heavier machinery needed to handle increased quantities of material in a given time. The above cost refers to bucket dredges. Dipper and suction dredges will cost less than bucket dredges; and, again, the suction dredge will cost less than the dipper dredge, in fact, a first-class suction dredge can be constructed for $15,000.

Mr. R. H. Postelthwaite claimed in 1897 that any ground not deeper than 60 feet below water level, or not more than 20 feet above, and which did not contain rocks heavier than 1 ton, could be handled at from 3 to 5 cents per cubic yard.

Subsequent practice has corroborated his statement, and in one case the cost has been as low as 2.36 cents per cubic yard.

The constituent parts of a dredge are:

(a) The hull, upon which the machinery floats.

(b) The excavator for digging and raising the material.

(c) The washing apparatus.

(d) The sluices and gold-saving devices.

(e) The tailings stacker.

(f) The power plant for driving the machinery.

The Bucket Dredge. — (a) **The hull** of all dredges must be substantially constructed and designed for the weight of the machinery it is to carry. More than one dredge has sunk because it was not properly calked or designed.

The construction of the hull is not a difficult matter and is carried on as for any lighter that carries its load on its deck.

The hulls are of wood and vary from 30 to 40 feet in width and from 60 to 120 feet in length. The depth of the hull is usually from 6 to 9 feet, and when completed and weighted with machinery draws from 4 to 6 feet of water according to the size. The frame of the scow is blunt pointed as in Fig. 53 for river dredging; but for inland dredging it may be constructed square at both ends, as there is no current to contend with. The bucket dredge has a well built in the bow that practically divides that in two parts. Fig. 54 is an elevation of the dredge shown in Fig. 53. Its different parts will be described in order as we proceed.

Fig. 55 is the dredge *Indiana*, built by the Bucyrus

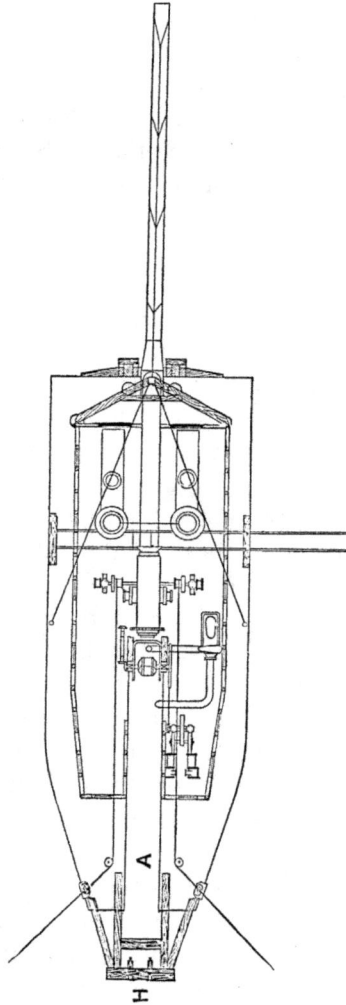

FIG. 53.

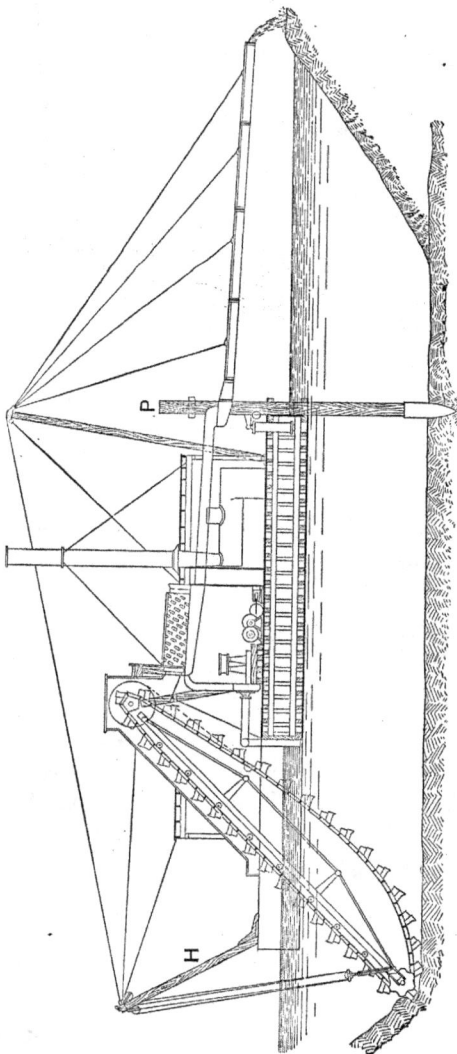

FIG. 54.

Company, and is used for the purpose of describing
the various essential details that enter into the construction. It is an electrically driven dredge, otherwise
boiler stacks would be in evidence as in Fig. 54. The
gantry, *a*, is constructed of heavy timbers, that rise
about 20 feet above the main deck. Rolled steel bars,
H-shaped, are sometimes substituted for timbers in
gantries, although they possess no particular advantages
over suitable timber sticks. The bucket ladder, *b*, is
a substantial steel-trussed arm, that extends some distance ahead of the dredge, the length depending upon
the depth to bed rock. It is pivoted to the driving
shaft at the upper end, and supported by a bail, *c*,
which is suspended by wire ropes from blocks attached
to the cross-tree of the gantry. The ropes are connected
with power, so that the ladder may be raised or lowered
as found necessary while excavating. At each end of
the ladder there are tumbler wheels, whose object is to
give motion to the buckets. The upper tumbler is
keyed to the shaft which carries the driving wheel, *d*.
Rollers are placed at intervals on the ladder, in order
to decrease the friction of loaded buckets, *e*, as they
move upwards to the dump, located on the top deck of
the boat. The buckets can dig in deep or shallow water,
but must be pitched in each case by the bucket ladder.
The tumbler under water is rotated by the bucket chain,
which is set in motion by the upper tumbler revolving
on its shaft.

(*b*) **Excavators for Dredges.** — There are three kinds
of excavators for dredges, — the bucket, dipper, and
suction pump.

FIG. 55.

1. The bucket excavator not only digs the dirt, but elevates it to a hopper at the top of the dredge. The bow end of the hull, as previously mentioned, is divided through the center in order to permit the bucket ladder to be raised and lowered, and the elevator buckets to travel. This kind of dredge is not tipped when a load is lifted from the river, consequently the plant is evenly balanced, even though the heavy machinery is placed near the stern. This type originated in New Zealand, and very little if anything has been added to its improvement since it was introduced in this country. The movement of the buckets is slow and uniform, the rate of travel being from 18 to 20 buckets per minute. The speed should be regulated to deliver the material in a fairly uniform manner, as the feeding is a matter of considerable importance. With buckets the material is brought up in comparatively small masses, that permits of it being properly washed without overpowering suitably designed screens.

Buckets are usually made of steel with lips reinforced by manganese steel strips varying in thickness from 1 to $1\frac{1}{2}$ inches. The lips are the weakest part of the bucket dredge, and if they were not reinforced as described the entire bucket would need replacing from time to time. Considerable wear and tear occurs when working in hard clay containing boulders, as a boulder in such material is not easily removed, and will cut and wear the bucket lip quickly, and often so dent the bucket even when reinforced as to make it practically useless for cutting a bank. To obviate this difficulty it is customary to drill holes ahead of the dredge working inland

and shake the ground with a blast. This will enable
the bucket to pick up the boulder or move it one side.
Here the Keystone driller again finds employment, in
drilling blast-holes ahead of the dredge.

The buckets are secured to chains by rivets, and are
with the chains made so strong that if they encounter
an obstruction that they are unable to move or to glance
from, they will stop the machinery. The capacity of
the buckets is from 3 to 13 cubic feet; the most usual
sizes being 3, 5, and 7½ feet, and these at a speed of
18 buckets per minute will theoretically deliver 120, 200,
and 300 cubic yards per hour.

Owing to the imperfect filling, the practical delivery
will average about two thirds of the above quantities.
The Bucyrus buckets are considered more efficient than
Risdon buckets, where there are no boulders, and the
Risdon buckets are considered to be better where there
are many boulders. The only difference between the
two consists in their hooking-up and chain construction.
The Bucyrus buckets are placed close together, while
the Risdon buckets have a bare chain link between them.
It is claimed by the Bucyrus people "that Robinson's
patent steel chain has advantages over all others, inas-
much as the chain pins are protected and lubricated so
that sand cannot cut out the links and pins, necessitating
their frequent renewal."

The depth to which the buckets may work is limited
by the power of the engine and length of the bucket
ladder, but for deep dredging the boat must be constructed accordingly. Probably the average depth below
water level so far worked by these machines is 40 feet,

although many have been constructed to dredge 60 feet.

The buckets deliver their contents into a hopper, so constructed that all material falls into it. No material should be allowed to return directly from the bucket into the water, as there is a probability of its containing gold. A stream of water should be used to clean the buckets at the hopper when digging in clayey or sticky ground.

The advantages of the bucket excavator are:

It delivers the material to the hopper in a comparatively uniform manner;

It can dig deeper than other excavators with less expenditure of power;

It requires but one hull, and all machinery can be placed on that hull.

The disadvantages of the bucket excavator are:

It cannot raise boulders out of the way, but bangs against them to the detriment of the buckets, without raising material, thereby requiring that the barge be moved;

The buckets are only partially filled, and permit some fine material to run back to bed rock, which they are unable to clean, if hard, and only partially to clean if soft.

2. **The Dipper Excavator.** — The dipper was first adopted for dredging gold in this country. It has been systematically decried, although in some ground it has no equal.

The dipper excavator, up to a certain depth, depending upon the length of the dipper arm, is able to move and remove larger boulders than the other excavators.

This is made possible by the fork on the end of the dipper, the large mouth of the dipper, and the concentration of power. These advantages become more evident when one takes into consideration that they obviate the necessity of raising and lowering the bucket ladder, backing and filling, with consequent cessation of work during that time, and swinging the scow in position. Further, it is possible to excavate more ground in a given time, closer to bed rock, and on account of the lateral swing of the bucket arm, a wider space without change of the scow's position.

The opponents of the dipper usually advance two arguments against its use: First, that the dipper door cannot be made tight without great expense for gaskets — and consequently there is apt to be a loss of gold unless they be used.

To obviate this loss from leakage, gaskets of common rubber hose are so arranged as not to come in contact with the material during its discharge from the dipper. These gaskets wear well, are not expensive, and can be replaced in ten minutes' time, if necessary, but they are not absolutely water tight.

As the material is brought up in masses, with little water in comparison to the bulk of material raised by the other two classes of excavators, the loss of gold from seepage through the door under any circumstances is slight, and probably not more than occurs from the continual stirring up and sliding back of the ground where buckets do the excavating, and which must necessarily loosen and precipitate some gold.

The second objection to the dipper is stated to be

FIG. 56.

the agitation caused by the dipper's attack on the material. This attack is no more vicious than that of a bucket in comparison with their sizes, the water in the dipper being pushed out gradually as the material enters; this objection is tenable only where the ground is loose, and the gold is free and very fine; however, in such instances the bucket is likewise objectionable.

The specific gravity of gold is such that particles the size of pin heads are not easily floated in swift-running water, and hence the approach of the dipper is not apt to cause the gold to float one side.

Mr. P. Wright says "that in the Beechwood district of Australia" he found 95 per cent of the gold within three feet of where it was filled into the sluice, the gold lying on a smooth board, and yet a powerful current failed to move it.

Mr. Alex. J. Bowie says that 80 per cent of the gold recovered is found within the first 200 feet of the sluice, and quotes an instance where in a 100 days' run which cleaned up $63,000, $85\frac{7}{10}$ per cent was caught in the first 150 feet.

Careful consideration of the imaginary difficulties attending the use of the dipper which are advanced will probably lead to its later adoption, since it has no equal for moderate depths and wide range for handling material.

A seriously objectionable feature of the dipper is the intermittent manner in which it brings gravel to the hopper; at times it delivers a full dipper, but more frequently a less quantity, with much water. It is difficult to receive material in this way, and generally the dipper will require another scow to treat the material, otherwise the hopper and sluices must be placed upon the river bank.

The same objections apply to the clam-shell bucket in a greater degree, for, should a stone prevent the shells from closing tight, the gold would be lost in getting the material to the hopper; besides, the agitation consequent upon the shutting and lowering of the dipper may be sufficient to float gold away from the material being excavated. These excavators work very satisfactorily in dry placer ground.

The dredge shown in the illustration was one of the first constructed in this country, and is, or at least was two years ago, in active operation on the Chestatee River, Georgia.

It will be noticed that the dipper is mounted on one barge and the sluice on another. This on first thought is objectionable, but on reflection there does not appear much doubt but that a barge constructed particularly for cleaning the gold furnishes a better opportunity and more space for gold-saving opportunities. Some dredges are constructed so that the bucket arm can be brought back to dump in a bow hopper placed in the center line of the barge but extending over the bow; others have hoppers constructed on the sides near the bow. In case that the hoppers are in the center line of the boat, spuds must be used in the rear; but where the hoppers are placed to one side or on a separate barge, spuds must be used at the rear and on the dumping side. Several dredges of the dipper type are working successfully on the Chestatee River, and the dippers for such dredges vary in capacity from 1 to 2½ cubic yards. The dredge illustrated washes the material through a grizzly by a stream of water into a sluice box, that discharges the tailings in the river. This dredge has considerable black sand to contend with. The dipper dredges at work at Oroville, California, use tailing stackers, and pumps for disposing of the tailings. They also use screens for disintegrating and washing the dirt. According to the Marion Steam Shovel Company, the two dredges in Oroville are among the most successful in that field to-day, although there are many bucket dredges there for comparison.

3. **Suction Dredges.** A centrifugal pump with a 12-inch suction hose that reaches to the bottom of the river, comprises the excavating arrangement on what are termed suction dredges. The hose is attached to a moving crane on the barge so that it can be moved as desired.

Mechanical devices and water jets have been proposed to loosen the material at the suction end. These, however, are not required ordinarily, as the pump will raise sand, gravel, and even rocks approximately the diameter of the hose in size.

The most successful dredging plant of this kind was that of Sweetser and Burroughs on the Snake River near Minedoka, Idaho. This plant consisted of a 15-inch centrifugal pump, directly connected to an 18-inch turbine water wheel. The gravel and water together were elevated a height of 25 feet, the suction being 20 feet, and the delivery pipe 5 feet. The material handled ranged from fine sand to 8-inch boulders. While the suction and discharge pipes did not wear rapidly, the pump propeller and casing could not be kept long in repair, and this necessitated frequent stoppages, until a steel lining and a closed impeller were introduced. Even with these improvements the results were not entirely satisfactory. In dredging on the float coal in the Susquehanna River near Nanticoke, Pennsylvania, there was very little wear on the pump impeller and casing; in fact, there was nothing in the way of repairs done to the pump in two seasons, although it handled large quantities of sand, coal, and rounded stones.

While the system is cheap as regards first cost, and takes up but little deck-room, it is very difficult to regulate. At one time the pump will be choked; at another time nothing but water will be delivered; furthermore, a natural selection of sizes takes place. At one time there will be a free flow, but the larger material that moves slowly will gradually form a layer that stops all flow. To keep the Susquehanna dredge up to its work, it required one man with a long pole to keep the material stirred up, and to move the tail pipe from the hole it quickly dug to a new position where the material was within the suction radius. The cost of working gravel on the Minedoka dredge was 2 cents per cubic yard. Its success was due to gravel being washed to the pump, to its working where the tail pipe was practically always in sight, and to its being run by water-power. The quantity of water pumped is always in excess of the material, and the power consumed is out of proportion to the material raised.

Centrifugal pumps revolving at high speeds are able to raise large quantities of water a short distance; and within the small radius of their suction, necessarily close to the suction pipe, they have sufficient power to raise mud, fine sand, and gravel.

This suction might raise considerable gold, because the velocity of the water entering the suction is greater than the velocity with which gold will fall by gravity through water.

The radius of suction is not sufficiently strong on its outer periphery to draw in heavy particles of gold to the central point of suction, while it is sufficiently strong to draw in sand and mud. When the sand and mud are disturbed the heavy particles of gold begin to settle and finally reach the bottom, in which position it is difficult for a suction pipe to dislodge them, particularly if there are boulders on the river bottom, or crevices into which the gold may sink. Suction pumps raise gravel stones without much difficulty, but they cannot raise coarse stones and boulders; and where the bottom rock is covered with these by natural selection, the gold falls in between them and is lost, unless bed rock can be cleaned by hand. As an auxiliary to other dredges where the bed rock is hard and not uneven, they might prove useful in cleaning up gold which the other excavators cannot recover. The suction pump as a gold dredger has a very narrow range of usefulness compared with the bucket and dipper. Another drawback to centrifugal pumps is the height to which they can deliver material above the pump. They usually are more successful at the suction than at the delivery end of the pipe line, and it requires excessive power to raise water and material higher than 12 feet.

(*c*) **Washing and Screening.** — Screens on dredges, if grizzlies are neglected, are either of the revolving or the shaking types.

1. The revolving screens are of sheet iron with holes ranging from .5 inch in diameter at the receiving end to 5 or 6 inches diameter near the discharge end. At present screens are made from 3 feet to 4.5 feet in diameter and from 20 feet to 30 feet long, according to the size of the buckets and capacity of the dredge. The diameter of the screens must be such that they will pass stones as large as the buckets will bring up. To prevent excessive wear on screens, it is a better plan to wash all material through grizzlies in the hopper and pass all material over 3 inches in diameter either to the tailings stacker or over the side of the boat. This will permit finer sizing of the material going to the sluices and tables, a matter as important in sluicing as in other kinds of concentration, where close recoveries are made. The length of the screen depends on the kind of gravel to be washed. In every case it should be such that no crowding of the material will occur and prevent it being thoroughly washed. Ground that contains water, rounded stones, and much clay, should be passed through rotary screens, as they are better washers and disintegrators than shaking screens. Jets of water, under pressure, issue from pipes inside the screen, if the screen revolves on rollers; or from the hollow shaft if the screen revolves on a shaft. This water washes the fine material from the stones and through the screen openings. In most cases there is but one screen; but two screens, one inside the other, are better adapted to sizing and

FIG. 57.

gold saving. The first screen discharges into the sluice, and the second one passes all material to a tank directly under it. The tank is a distributing box that discharges its contents uniformly over the gold-saving tables. In case there is but one screen, the large stones discarded at the end go to the tailings stacker, while the material passing through the screen openings goes to the sluice, and the fine material to the gold-saving tables.

When a single screen is used the size of the openings must be such that no gold will be lost. This is a very uncertain factor, consequently there is need of some large openings, and the necessity of turning some material into the sluice box. With a double screen the lower screen may be rotated in water and only fine stuff sent to the tables, while the discharge may be delivered to the sluice. Screens should be constructed in sections, in order that a worn part may be replaced without discarding the entire screen.

The description furnished of a rotary screen arrangement is the author's ideal. As now constructed, the material that passes through the screen in most cases passes directly to a sluice box containing riffles and not over a gold-saving table, the supposition being that only coarse gold is in the placer.

2. **Shaking Screens.** — It has been stated that rotary screens are better for clayey material and round stones than shaking screens. Clean gravel with little clay, and material carrying fine gold, are suited to shaking screens. Rough stones are handled with equal facility by either kind of screen.

Shaking screens are given sufficient fall and screening area to permit of the gold being thoroughly washed from the rocks before the latter are discharged to the tailings stacker. Fig. 57 shows a shaking screen with gold tables. In the illustration the shaking screen, which has an area of from 600 to 700 square feet, works in the box, *a*.

From a series of openings above the screen, water is projected upon all parts of the screen, washing and disintegrating the material before the finer particles pass into the chute, *b*, leading to the distributor, *c*, placed below the screens and above the gold-saving tables. *d*. The coarse material passes out at the lower end of the screen into a hopper, *e*, leading to the tailings stacker.

The fine material which passes over the tables is carried to sluice boxes, which discharge the material about 20 feet beyond the stern of the dredge. The sluice boxes are also fitted with riffles, and if there is much sand they deliver their material to a centrifugal pump for final distribution. In this case the pump discharge pipe is placed on the stacker arm so that the sand is delivered over the top of the gravel (see Fig. 55). Similar arrangements may be necessary with a rotary screen, if gold tables are used in addition to sluice boxes.

(*d*) **Gold-saving Arrangements.** — Before an intelligent conception can be reached of the kind of gold-saving appliances to adopt, the fineness of the gold going to the riffles must be determined. The next important consideration is the material, and the best arrangements to install to wash it thoroughly. As too much water is almost as bad as too little water, experiments should be made to ascertain the minimum quantity of water required for a maximum recovery. Any neglect of one of these three important items may make the sluice or table riffles but mediocre gold-saving appliances.

Hungarian riffles or gold tables such as are shown in

FIG. 58.

Fig. 58 are used on dredges where coarse material is separated from the fine material before the latter passes over the tables. They require considerable mercury, and are fairly effective when given a grade of about 18 inches in 12 feet. They are placed below the sluice-box grizzlies or distribution boxes, and are virtually undercurrents. The grizzlies do not allow anything larger than ½ inch diameter to go to these tables.

The riffles most generally used are of the Hungarian type as previously illustrated, but are supplemented

by the riffle shown in Fig. 59, which is made up as follows: First an iron floor upon which is placed a layer
of calico. Above this in the order named is a layer of
ordinary cocoa matting, and expanded metal. The metal
is fastened in such a manner it can readily be removed
when it is desired to wash the cloths. Every few days
the expanded metal, which is used to keep the matting

FIG. 59.

flat and hold it down, is taken up, and the cloths washed
in a box to collect the fine gold, after which they are
returned to the tables, and the expanded metal fastened
in place. These tables are considered the best fine gold-
saving devices, and are widely used in consequence.

In case there is much black sand, the matting becomes
clogged so that the gold cannot settle, and if more water
is run over the tables to clean them of sand, the gold
goes with the sand. To prevent the accumulation of
black sand it must be removed before it reaches the

matting. Owing to the limited length of sluice boxes on dredges and their narrow width, undercurrents or gold-saving tables are made wide; on some dredges they occupy as much as 1200 square feet of space. There are no better gold-saving devices in existence than are found on dredges, consequently the loss must be due to some extent to imperfect washing arrangements, or to allowing too much material to flow over the table at one time. A steady flow of water is imperative in such work, even should there be an uneven flow of material.

If the water passing over the tables contains much alumina or much magnesia, it will slick the tables in a very short time and prevent anything but heavy gold adhering. Recognizing the necessity of thoroughly washing the gold, and at the same time comminuting adhering substances to such an extent that they would be held in suspension by the water, the writer suggested in 1897 that the log washer be adopted on dredges. The success attained by the log washer in the South has verified the author's expectations. In some cases the screens on dredges have been arranged to pass all fine material to centrifugal pumps for additional washing, and these pumps have delivered the washed material to the sluices. There are several objections to this method of washing: First, there is too much water delivered to the sluices; second, the centrifugal pump is a poor washing contrivance; third, there is an unnecessary waste of power; fourth, the material is necessarily delivered in an intermittent manner to the pump and to the sluices, consequently more water than is needed must be pumped. Sluice boxes on dredges are about of the same construction as in ordinary sluicing, except that they are often of iron. Where the rotary screen is used without gold-saving tables the sluice boxes are made long, as shown in Fig. 53, and in some instances they extend back to independent barges, that support them in such a way that their length can be materially increased, and an undercurrent placed in the sluice line.

(*e*) **The Stacker.**—There are two kinds of rock stackers in use, both of which are of the endless belt type. The stacker arm is steel trussed, and is raised or lowered as occasion demands by wire ropes that are worked by a small hoisting engine. The ropes are fastened to the bail of the stacker arm and pass through pulleys suspended on the cross piece of the stern gantry.

At the top and bottom of the stacker arm there are sprocket wheels or belt pulleys; the lower one in most cases is the driver, while the upper one is the driven wheel or pulley. This arrangement is not as economical in the use of power, or as satisfactory, as where the upper pulley wheel is the driver. The rock stacker is particularly useful in ground where there are many stones, and in places as much as 75 per cent of the material is coarse. It also aids in saving gold by keeping stones out of the sluice boxes, and in this way shortening their length.

The bucket conveyor on the tailings ladder is said to last longer than the belt conveyor, and to raise material at a higher angle. It requires, however, more power to run, is more expensive in first cost, and less easily repaired, than the belt conveyor. Belt conveyors have carried 500,000 tons of sharp rock without wearing out or needing repairs; in fact, they are preferable to chain conveyors where the angle of elevation is not so high as to cause the material to run back, or about 24 degrees. The tailings stacker must be long enough to raise the coarse material to an elevation and to a distance behind the dredge that it will not run back into the diggings.

When fine material is to be put out of the way, the pump pipe is attached to the tailings ladder, and the pump must have sufficient power to throw the sand and water over the top of the coarse tailing pile. When the pump is used to dispose of the tailing the tail sluice is not needed, as all material that passes down the sluice box goes to the pump.

(*f*) **The Power Plant.**—Electrical power, when it may be obtained readily and cheaply, is preferred to steam-power. Where two or more dredges are in close proximity, it is more economical to locate a steam boiler plant on shore and transmit electricity to the dredges, than to have engines and boilers on the boats. It takes considerable handling to place fuel on board the boat, and it is a well-known fact that several small engines are wasteful of steam to a greater extent than a large automatic cut-off engine when generating electricity.

There will also be a loss of steam due to condensation, on a dredge, as the engines must be placed at some distance from the boilers. In addition to the disadvantages named, the boiler is in the way on the boat, unless additional length is added for its accommodation. In localities where there are ditch lines, electrical generating plants can be cheaply installed by using nozzles and impulse water wheels. In other localities a small dam and flume can be cheaply constructed to furnish water-power for generating electricity.

The power required for a 3 cubic foot bucket dredge is about as follows:

For driving the buckets . . . . . .   75 horse-power.
To drive a 10-inch centrifugal sluice
        pump   . . . .   50 horse-power.
The revolving screen requires   . . .   20 horse-power.
To drive an 8-inch centrifugal screen
        pump . . . .   30 horse-power.
To move the scow . . . .   20 horse-power.

Auxiliary pumps, power for tailings stacker, electric light plant, and sluice pump for tailings, would require considerable additional power.

In the above estimation, which does not include the power required for a tailings stacker or sand pump, it will be observed that 80 horse-power or 41 per cent of the power is used for washing purposes, and but 38 per cent for digging purposes.

The object in using centrifugal pumps is to obtain a large supply of water, and this is accomplished at the expense of power. The pumps are cheap and easily kept in repair; however, it is probable that by using compound centrifugal pumps, the power would be economized particularly in the spraying pumps, where force is desired rather than quantity. A reduction in the quantity of water used in sluicing would often be found beneficial. The object is to transport the material and not flush the sluice boxes.

The following table furnished Mr. D'Arcy Weatherbee

by D. P. Cameron of the Western Engineering and Construction Company, who are agents for the Bucyrus Company, gives an idea of the weight of a 3½-foot Bucyrus dredge.

Name of Part.	Total Weight Lbs.	Number of Pieces and their Weight.
Upper tumbler	6,500	Can be cut in 20 pieces, one of which will weigh 1,000 lb., the rest will be below 300 lb.
Lower tumbler	4,500	Can be cut in 13 pieces, three of which will be about 700 lb., the rest below 300 lb.
Digging ladder	28,000	Two pieces of 600 lb., the rest about 300 lb.
Digging buckets (3½ ft.).	83,000	Bottom about 320 lb., each hood 135 lb., lip 120 lb.
Screen, stacker and parts	16,000	Eight pieces would weigh about 600 lb. each, all other parts 350 lb. and less; 70 per cent less than 300 lb.
Gearing	30,000	Eight parts would weigh about 700 lb. each, the rest from 350 lb. down; 50 per cent less than 300 lb.
Engine or motors . . .	15,000	Two pieces about 1,000 lb.; two pieces about 600 lb.; 50 per cent below 350 lb.
Boilers	8,500	All below 350 lb.
Pumps	300	
Winches	42,000	Two pieces 600 lb. All other parts below 350 lb.
Other parts	7,600	All below 350 lb.

Spuds. — On the bucket dredge there are two spuds 42 × 18 inches × 50 feet long with steel points at the lower end. The spuds, which are raised by machinery and lowered by gravity, serve to move the boat, or hold it steady when dredging. To move the boat forward or backward the spuds are alternately raised and dropped, after the engineer swings the boat by means of cables passing around the front corners of the boat and attached to lateral anchorages.

When dredging, one of the spuds rests on the bottom and forms a pivot, around which the boat is swung as the gravel is taken up. The buckets thus take off a segment of dirt about 6 inches deep and 8 feet wide, and after each swing of the dredge around the spud the ladder is lowered 6 inches. The lowering of the ladder continues until bed rock is reached. The bed rock, if yielding, is torn loose and brought up until barren of gold.

The dipper dredge is supplied with 4 spuds, one near each corner to prevent the barge from swinging and from tipping. The spuds have racks, and are raised by pinions driven by machinery.

The boat is moved forward by ropes attached to anchors and winches. The boom swings 180 degrees; consequently the dipper can dig quite a semicircle, and to a depth depending on the length of the dipper arm, without changing the position of the boat.

CHAPTER XI.

TRACTION DREDGES: DRY PLACER MINING MACHINES.

STEAM shovel excavators have been mentioned under the caption " Exploiting Placers." In that connection, however, there was sufficient water for the excavation, but a lack of dumping ground. Traction dredges are for exploiting placers where little water exists, and where conditions are unfavorable for sluicing, dredging, or the use of other systems of placer mining.

To determine whether this method of work would be profitable, exploration and prospecting must be carefully carried on. Sure thing placers do not exist in all localities as they do in Oroville and some other districts, therefore where one test hole was put down in every four or five acres, one and probably more holes will be required for every acre. The land is therefore divided up into sections, and in some cases a Keystone drilling machine takes samples just ahead of the dredger in order to work the richest ground. When communicating with manufacturers of traction dredges the following information in detail is required by them:

The lay of the ground; that is, whether it is in a gulch, an old river bed, lake bed, or small valley;

The grade of the bed rock if that can be determined, and if not, the slope of the surface;

How high the material must be raised in order to obtain sufficient sluice fall;

The kind of material to be washed, *i.e.*, whether coarse or fine;

The depth from the surface and the thickness of the pay streak; this will furnish practical information regarding the quantity of waste material that must be handled and disposed of;

The quantity of water at command and the distance it must be piped;

Water in cut, if any, and how deep;

A contour map of the ground is very desirable.

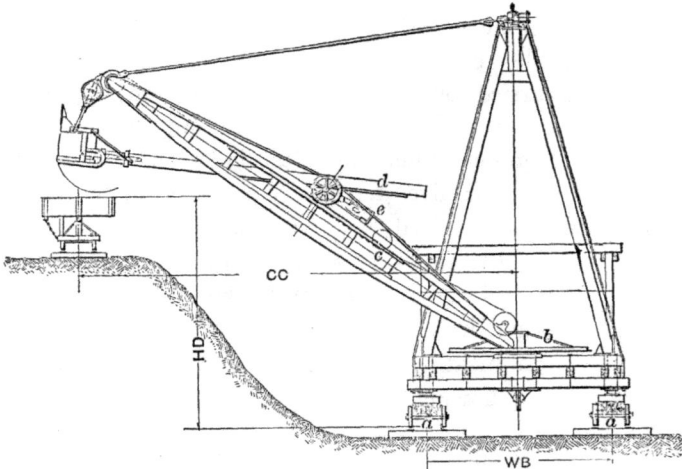

Fig. 60.

The end outlines of a traction dredge with plain swinging circle are shown in Fig. 60.

The dredge platform rests on two trucks, *a*, that

have a 27-inch gauge and are moved by the motive
power used to drive the other machinery. The distance, WB, from center to center of the tracks is from
12 to 14 feet. The circle, b, is for swinging the boom, c.
The length of the boom required depends upon the
height of the dump, or HD, above the dredge track,
and the distance, CC, from the center of the machine

FIG. 61.

to the center of the dump. The length of the dipper
arm, d, depends upon the depth of the cut and the
height, HD. The machinery on the dredge when the
dredge is not self-contained consists of an engine for
working the boom, a thrust engine, e, on the boom, and a
boiler. The capacity of the dredge is governed by the

power of the digging apparatus and the size of the dipper.

Self-contained dredges in addition to the machinery mentioned will require power for hoisting the car to the dump, revolving the screen, and working the tailings stacker.

Fig. 61 is one of a number of traction dredges constructed by the Marion Steam Shovel Company. It is a rear view. The skip, *a*, is loaded on the bank by the shovel, *b*, and is then hoisted and dumped automatically into a hopper. The ore is washed into screen, *c*, the fine ore going to the sluice, *d*; that which passes out of the end of the screen to tailings stacker, *e*; and the very coarse goes over the side of the dredge.

The platform rests on four trucks. The machinery is so arranged that it is not crowded, and comes within the center of gravity of the car platform, thereby doing away with jackspuds and braces, which are necessary when there is but a single track and a narrow platform. The platform sills are of wood or steel girders, stiffened by ties of iron, forming a king or queen truss extending the entire length of the sills.

The trucks, platform, and machinery will weigh between 40 and 70 tons.

Single-track traction dredges have been constructed to run on a 4-foot 8½-inch standard railroad gauge, in order that all parts might be assembled at the shops and the machines transported to their destination. There is not much gained by this construction, from the fact that it is not often that placers suitable for this method of exploitation are found near railroads; and to lay a temporary railroad to the placer ground is expensive, even if the dredge contains its own motive power.

Where single-track dredges are used, jack arms and side braces must be adopted in order to keep the machines upright, and prevent the dipper when swinging from straining the parts of the car body.

The tracks for traction dredges must be kept as near bed rock as possible, and at the same time the machinery should be kept level, to prevent undue wear on the journals as well as keep the water in the boiler in proper position. These machines are said to do work on considerable incline, but they are not built for that purpose, and will save money for the operator if kept level. The trouble with the first machines of this dry-placer type was that they cost as much to keep in repair as the value of the gold saved, and, as they were discarded, probably more.

Mining with such machines will depend upon the water supply. Beside a river bank or near some stream they should work satisfactorily, but in situations where water is not abundant they must be economical in its use. If it be necessary, 85 per cent of the water needed for working these machines may be impounded and used over again, thus requiring but 15 per cent of the total quantity to be fresh.

The water supply must in all cases be in quantity from 8 to 10 times the amount of dirt excavated.

Thus, if one cubic foot of dirt be washed per minute, there will be required from 8 to 10 cubic feet of water needed per minute; of this amount from 6.8 to 8.5 cubic feet may be used over; thus the actual fresh supply required will be from 1.2 to 1.5 cubic feet per minute.

With first-class washers the amount of water required

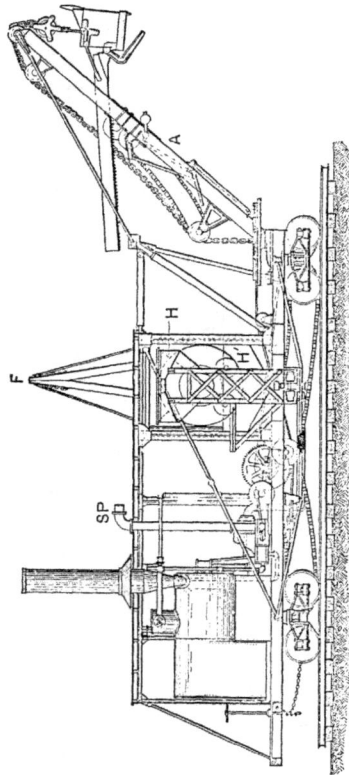

FIG. 62.

should not be more than 8 cubic feet per minute.

The dirt is excavated by an ordinary steam shovel whose dipper is capable of handling hard pan and ordinary hard material, or by the clam shell bucket. The dipper of the shovel works from the arm of a derrick, so arranged in this instance as to have an arm long enough to deliver the material directly over the hopper, H, Fig. 62. The derrick is mounted on a turntable which is made to revolve by machinery nearly 140 degrees, or until the dipper is directly over the hopper.

The dipper, being required to excavate hard cemented material, must combine strength and power. The boom for the bucket arm is made to conform to the depth of the alluvions. For example, a 35-foot boom will raise material 18 to 20 feet above the track and make a cut 35 feet in width.

With the exceptions of the length of arm and the turn, the excavating part of the machine differs very little from the ordinary railroad steam shovel. Where the washing machinery is on trucks at the back or at the side of the shovel, the swing may be halfway round. In some instances the shovel is independent of the washing machine, the latter being stationary and the shovel only advancing. Where the washer is stationary, tram cars or traveling conveyors are used to carry the material from the shovel to the washer. Dippers of the scoop shape are generally used, although clam shell buckets will answer in some cases. Scoop dippers made to hold $1\frac{1}{4}$ cubic yards will when filled probably not average over 1 cubic yard of dirt. They could under favorable conditions make six scoops and deliver six buckets into the hopper in five minutes, or 72 cubic yards per hour; however, at this rate, under ordinary circumstances, the washer could not handle the material, consequently 1 cubic yard per minute should be assumed for calculations. Where there is plenty of water the shovels can be increased in size up to 2½ cubic yards, but the whole plant must necessarily be enlarged in proportion.

Wherever the hopper for the reception of the excavated material projects beyond the side of the car it must be strongly braced; further, the structure is subjected to considerable vibration and strain by the sudden unloading of a cubic yard of material. Another disadvantage is that the hoppers require too much fall for the height of the machine, necessitating the use of power in raising the waste material to the dump and the pulp to the sluices. To avoid the strain from side hoppers, some makers place the washing and elevating apparatus upon separate cars. It is possible by the use of a wide platform and the double truck system mentioned to raise the washing machinery and allow gravity to dispose of the coarse, medium, and fine material without recourse to elevating machinery for that purpose.

To accomplish this the washer is constructed on the car platform and the hopper placed for the reception of excavated material above the washer but within the center of gravity of the car.

Another system of raising the material to the hopper is where a double inclined track is laid from the ground to the top of the mill. Upon this track two skips run; as the loaded skip ascends, the empty skip descends. The power for raising the loaded skip is derived from the engines which work the excavator. The material having been dumped automatically into the hopper, it is washed down over coarse screen bars.

That portion of the material too coarse to pass the bars goes directly to the dump by gravity; that portion which passes the grizzlies falls into the screen, where it is thoroughly washed of fine material, which falls into the sluices, while that portion too coarse for the sluices moves by gravity to the dump. This system disposes of all tailings and pulp by gravity, thus making an economical and power-saving system, by doing away with elevator engines and one pump, as well as the elevating and conveying apparatus.

The hoppers in dry placer mining machines should be so arranged that the material may be washed by water from pipes, P, surrounding the hopper, and through iron bars forming the floor of the hopper. This will allow the action of the screen to more thoroughly disintegrate the material. The coarse stuff remaining on the bars can be removed by mechanism down over a stone chute. The screen should be of two compartments. The inner compartment (being fed by streams of water to further soften and wash the material) should allow the passage of all stuff up to $\frac{1}{2}$ inch diameter into the outer compartment. This outer screen should be arranged to revolve in water, thus further washing and disintegrating the material. The pulp from the washing hopper is drawn off by a centrifugal pump and raised to the sluices containing the riffles.

The coarse stuff from the inner circle of the revolving

screen falls into elevators at *B*, Fig. 63, and is conveyed by them to the dump.

In the illustration, Fig. 63, which is the Traction Dredge of the Bucyrus Company, the hopper is supplied with water from pipe, *P*, which washes the material down

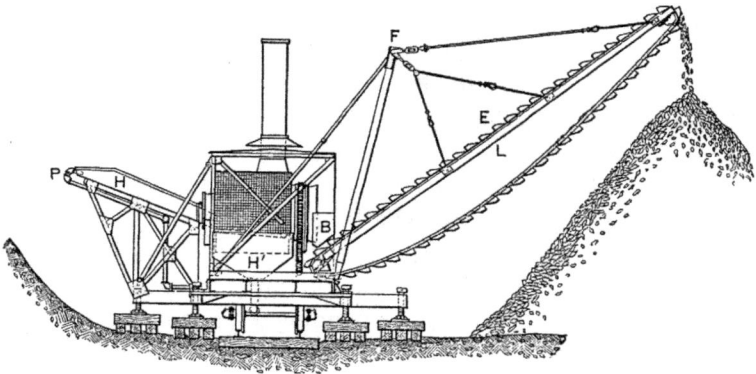

FIG. 63.

into the screen; a second hopper, *H'*, receives the washed material containing the gold. The pipe, *SP*, Fig. 62, is the pipe for discharging the pulp into the sluice box from the pump. *F* is the A-shaped head frame which supports the bucket ladder, *L*, over which the loaded tailing buckets travel from the screen discharge to the coarse tailing dump.

The sluice boxes are not shown. They may be extended a considerable distance from the machine, but if water is scarce the material is discharged where the water may drain into a sump. With plenty of water a one per cent grade will carry off the material in the sluices, which are provided with riffles. The first few sections of the sluice box should be of light steel, so that they may be readily handled and made water tight.

The Chicago Mining Machine has a complicated screening arrangement, and a short riffle sluice on the machine itself. The tailings from the riffle sluice are discharged upon the coarse tailings dump. This company pays particular attention to washing the material in the revolving screen, which has in its inner compartment a spiral conveyor. No pitch at all is given to the screen, the material being moved forward by the conveyors.

The list of machinery for such dry placer machines comprises a boiler of the upright or locomotive type, engines to work the shovel and derrick, engines to run the washer and conveying machinery, pumps to supply the water to the washer and sluices.

The horse-power necessary to work the shovel is furnished by a double 8 × 10-inch engine, and may be rated at 25 H.P. To run the elevating and washing machinery 6 × 6-inch double engines are used, which may be rated at 10 H.P. Centrifugal pumps are used, and they will require 15 H.P. each for their independent engines. At times an auxiliary steam pump may be required, and in some instances it is part of the system to use it for pumping water to the hopper and washer, leaving the centrifugal pump to work the pulp only. The screens, elevators, sprockets, chains, rollers, etc., will vary in style and make, according to the machine manufacturers' patterns, and are therefore not described.

With traction dredges whose rated capacity is 1 cubic yard per minute, it is safe to estimate that in one hour out of every ten the machine must be stopped for repairs, or for advancing, or other cause, which will place the average duty at 500 cubic yards per day. The fuel will generally be wood, at $4.50 per cord, and two cords daily, or $9, for 50-H.P. engines. Wear and tear, oil and waste, will amount to 3 cents per yard, or $15 per day. The labor of 5 men, averaging $3 per day each, $15, making the total expenses of running such a plant, not including quicksilver lost, $40, or 8 cents per cubic yard.

This estimate of running expenses does not include the superintendent and his expenses, or the transportation of the gold dust. The latter two items will amount tc $10 daily at least, bringing the cost to 10 cents per cubic yard.

The amount of gold collected will depend upon the machine construction and the superintendent; a poor machine will not aid a good superintendent. Suppose a machine weighs 50 tons, or 100,000 pounds; the cost at the mine will approximate 7 cents per pound, unless some patents in connection with it raise it considerably higher. Suppose the value of the gravel is 20 cents per cubic yard, and 90 per cent of the value is recovered. The profit under the conditions cited would be $5000 the first year if the entire year could be worked through. There is no doubt but that traction dredges are better calculated for some conditions than other methods of washing gold, and that they have not received more general attention is due to the exploitation of floating dredges, and the unwillingness of operators in this line to experiment with anything new.

What has been said previously regarding the thorough exploration of placer deposits applies here. The location of the deposit with reference to the nearest railroad station, and the condition of the roads leading to it for transporting machinery, are matters of importance. In case it is impossible to transport the boiler, power may possibly be transmitted by electric wires from a distance.

Several reliable steam shovel concerns furnish the machinery and plans for traction dredges. These companies are not willing to build machines for placer work unless they are assured beforehand, by examination and thorough exploration of their own or some other reliable engineer, that the diggings are of sufficient value to make the enterprise a success. The Bucyrus and Marion Steam Shovel companies state this.

From the very nature of placer mines — that is, the cemented state of the gravel — it follows that if the material can be broken up before it reaches the sluices or the dipper the chances for gold recovery are improved. There are many instances where the ground is so tenacious or the banks so high that it is thought advisable to run in tunnels and counters to break it up with powder.

Experience in breaking down gravel banks with powder will satisfy most people that small blasts on the edges of a bank are more economical in the use of powder and more effectual in breaking material fine than large blasts in tunnels. For shovel work, a blast which merely jars the surface and does not throw out the material will afford easy working for the dipper, and, what is more essential, will permit the ground to be washed much easier. The effect of the shot seems to be that of rending the whole mass of dirt without displacement, hence it is very advantageous where water is scarce and steam shovels are used. If the dipper delivers large lumps of cemented gravel of a tenacious character to the hopper, considerable water must be used to wash it down so fine that it will disintegrate readily; but water in such cases is an item, and consequently any method which will bring the material to the hopper in such shape as to reduce the quantity of water to a minimum will help the washing and recovery that much, and further increase the capacity of the machine.

Small blasts are considered to require more powder than large blasts in comparison with the proportion of the ground they disturb. This is true to a certain extent, but it must be borne in mind that the ground is more thoroughly rended by small blasts than by large ones, and it is the results in detail which are sought; in other words, the quality rather than the quantity for traction dredges.

Dry Placer Machines are those constructed to work without water, consequently they cannot be as effectual as machines using water. There are many placers in Nevada, Arizona, New Mexico, and Lower California, where water is lacking, and in such placers all kinds of schemes have been exploited; and it may be set down as an axiom that all dry placer machines will prove failures, unless gold is so plentiful it may be sifted from the dirt, and under the latter conditions a 10-mesh sieve

would suffice.   There are cases on record where material is pulverized to some extent and tossed.   When
thrown up the wind blows the lighter material away,
while the heavier material is caught on a sheet.   This
is again tossed until the heavier particles are concentrated to small bulk and the gold picked out, or the concentrates carried to a place where water can be obtained.

Fig. 64.

The Allis-Chalmers Company exploit the Wood dry
placer machine, but the writer not being particularly
interested in that kind of mining has never inquired
into its virtues or where it has been successfully used,
unless the machine illustrated in Fig. 64, a description

of which is kindly furnished us by George W. Parker, represents that machine.

In the *Engineering and Mining Journal*,[1] 1903, a description of the Edison and Freid dry concentrators may be found. Both machines depend upon gravity and an air current to separate the lighter material from the gold.

The process is to size the material and send the sizes to separators adjusted to the size.

Dry washing is carried on at the Sunnyside mine near Round Mountain, Nevada.

An idea of the work performed can be derived from Fig. 64. The few large rocks are picked out by hand and the gravel thrown by shovels against a 1-inch sand screen. The screened material is shoveled into the dry washers. The dry washer consists of a screen with $\frac{1}{4}$-inch openings, from which the oversize is delivered by a piece of sheet iron 2 feet beyond the end of the machine. The undersize returns to the head end, where it is fed on a frame covered with a coarse heavy cloth, across which are riffles about 4 inches apart. The frame with riffles is shown in the foreground to the right, and the washer is directly back of the man cleaning the riffles. The frame when in place forms the upper side of a bellows that is turned by a crank having a flywheel. The puffs of air through the cloth agitate the gravel, and, aided by the slope of the frame, it is discharged at the lower end, while the heavier gold is retained in the riffles. The gravel and gold retained by the riffles are brushed off into a tub, and after a sufficient quantity

[1] Mr. George W. Packard, Mining Engineer, Boston, Massachusetts.

of this concentrate has accumulated, it is put over the machine a second time. The tailing from this second concentration contains gold and is sacked for shipment. The concentrate from the second operation is washed in an ordinary gold pan, and the black iron sands and gold separated by a magnet. Two machines working 10½ hours per day, handle 35 tons of dirt. It requires 20 men to dig this amount of dirt, screen and put it through the dry washers.

CHAPTER XII.

NEARLY every placer deposit contains more or less magnetite, meccanite, or ilmenite, and at times other heavy minerals such as garnets, platinum, and the platinum metals, monozite, etc. The origin of the sands is not difficult to fathom, for some of them are found in the tailings from stamp mills, which indicates that they were originally associated with other minerals in rock formations. Minerals of this description are not easily oxidized, and in some cases are not affected by weak mineral solutions, or acids. Black sands in some cases are so abundant, that they interfere with sluicing operations, particularly in some river operations. In most instances they carry gold, which varies from $\frac{1}{2}$ to 90 ounces per ton of clean sands. The latter quantity is not usual, however, but as high as 4 ounces per ton is not unusual.

The Minister of Mines of British Columbia publishes in his 1904 Report the value of some of the black sands in one sample at least from the Caribou District. The assay value was as follows:

Gold,	95 ozs. per ton.	Value $1900 per ton.	
Silver,	180 " " "	"	90 " "
Platinum,	64 " " "	"	832 " "

Palladium, 61.4 ozs. per ton. Value $1769 per ton
Osmiridium, 42 " " " " 1386 " "

The remarkable feature about this deposit is that the quantity of silver is far in excess of the amount usually found with placer gold; further, that the assay reflected some copper which was probably alloyed with the silver, although it may have been alloyed with the platinum.

The quantity of iron in the sample was neglected; however, all indications point to the iron sands having an attraction for gold when in solutions. The nature of the gold shows it to be in a very thin film about the oxides, as if placed there by solutions, and only solvents can separate the two. Cyanide solutions are quite effective in obtaining gold from black sands.

It will be found that the black sands on seashores are not as rich in gold as the fresh water sands of inland placers.

While there is no question in regard to the value of some black sands in placers, there is great uncertainty as to their quantity.

This uncertainty in placer mining is sometimes the cause of cocoa matting tables becoming quickly filled, but this would suggest a means whereby they could be accumulated as a by-product. Mr. John M. Nicol[1] writes interestingly on this subject, and we have therefore taken the liberty of inserting some of his ideas, which are pointed.

"Unfortunately for the interest of abstract science,

[1] *Mining and Scientific Press*, Jan. 19, 1907.

many other workers, like myself, are no doubt engineers employed by large firms, who cannot, in justice to themselves or their employees, reveal all the knowledge that they acquire. There are, however, a number of points of general interest open to discussion by all parties, and I take pleasure in calling attention to some of them.

" There is no new discovery about ' black sands.' They are to be found disseminated throughout the sand and gravel — both ancient and modern — of practically all river, lake, and sea-beach deposits. The various minerals are associated together by virtue of the fact that they are all of high specific gravity, and also, that owing to their durable structure, Nature has had the opportunity of concentrating them in the river channels from a vast area of country from which they were originally eroded, and the nature of the minerals composing the grains will therefore largely depend upon the geological features of this area. The grains are of all sizes, from a maximum of about $\frac{1}{8}$ in. diameter down to material that will pass through a 200-mesh screen or even finer.

" The whole question is simply one of ordinary placer mining, with some definite system worked out for saving a larger percentage of fine gold and also of saving all of the other by-products that have hitherto been allowed to run to waste, and of doing this on a strictly commercial basis, without undue capital expenditure and at such low cost of operation that a sufficient profit will result to repay capital, interest, and a good surplus besides, before the deposit is exhausted.

" Wild statements have been made regarding the value

of black sands, and samples have been submitted which assayed $1,000 per ton, but it must be remembered that these samples consisted of a few pounds *actual weight* containing a few dollars *actual value*, that had been concentrated down from possibly many hundred cubic yards of gravel, and that before one ton could be obtained probably 4,000 to 5,000 tons of gravel had to be washed down, so that the real value of the original deposit in place possibly did not exceed 20 cents per ton. The first thing to be done, therefore, is to base all reports on the value per original ton or cubic yard of gravel in place, from which the black sand concentrate has been obtained.

" In river deposits, the possible flood line is of great importance. Possible hydro-electric power sites in the neighborhood should also be noted, as there are many modern methods applicable, that were not within reach of the early placer miners, where electric power can be obtained conveniently for pumping, elevating, and conveying, and for driving the necessary machinery for any plant that may be installed.

" The ground should be thoroughly tested either by drilling or by shafts; if the seepage is not excessive, the latter is preferable. In case of an elevator proposition, the proportion of fine to coarse gravel must be carefully noted as follows :

" The proportion by volume per cubic yard of all gravel below 2 inches, from 2 to 5 inches and from 5 to 12 inches. Gravel over 12 inches must be considered as too large either for practical dredging or for hydraulic elevating.

" If conditions permit, the whole of the gravel extracted

should be washed, and the coarse gravel stacked on one side. This may be conveniently done by means of a washing platform discharging by a short sluice into two or more pairs of rockers, which may alternately be cleaned up at short intervals. No riffles should be put in the sluice, nor quicksilver used. The clean-up of the rocker riffles must be passed through a 10-mesh screen, the undersize going directly to a settling vat for further test treatment, and the oversize being roughly picked over and thrown to waste. Any coarse nuggets or gold that may be caught on the rocker grizzlies or on the washing platform should be kept separate and their individual weights and measurements recorded.

"That which is saved in the riffles of the rocker may be considered as material that could be saved by ordinary sluice-box methods and that could certainly be saved by sizing and concentrating on the tables. The tailing from the rockers must also be weighed and passed through a 10-mesh screen. The oversize should be roughly picked over and then thrown to waste, and the undersize 'panned down' by panning back and forth between two miner's pans. If skill and care are exercised, practically all of the black sand that has escaped the rockers can be saved by this method. All of the concentrate caught by this means, will have to be tested as mentioned below. •

"The gravels of different mines vary so greatly, as do also the proportions of the different associated minerals and the size of the grains, that it is more than probable that each particular plant will have to be designed with regard to the local conditions. I think, however, it will be generally conceded, that all gold and platinum grains above $\frac{1}{8}$ inch diameter are comparatively easy to save. The real question is how to save the fine, and especially how to separate as well as save the associated by-products, without destroying one to save the other. The following additional tests are therefore necessary, and will aid in throwing some light on a possible solution of this problem.

"The material saved from the settling vat of the rockers, and from the pan concentrate, should be carefully dried in sheet-iron trays over a roughly constructed furnace. For small tests, the miner's gold pan may be used for this purpose. Care should be taken to make certain that the settling vats are thoroughly clean, and also that the trays are carefully dusted after the product has been dried, as otherwise some of the fine gold is liable to be lost. The dried product should now be weighed in bulk and a record made of its actual weight in pounds and of its proportionate weight per cubic yard of gravel from which the concentrate was extracted. For small hand tests, it is most convenient to give the weight in grams, and for the large tests in pounds, though if the facilities exist, it is far best to use the metric system right through. Unfortunately most miners do not understand the metric system, and a report to be intelligible to them, must be given in cents per cubic yard, and the weights in pounds.

" The dry product should now be passed through a series of ordinary laboratory sieves from 10 down to 100 mesh or even finer. Each oversize will be carefully separated and weighed in bulk, the magnetizable product will be removed and weighed, and the residue can then be conveniently treated by the old-time method of blowing, care being taken to blow the tailing on a large sheet of paper, so that it can be collected for further treatment. The gold particles remaining in the small blower must then be weighed on a button balance and placed in a small vial properly labeled for future reference. An assay sample should be taken from the original pan concentrate and also from the magnetizable product and the residue after the gold grains have been removed, and assays made and recorded. This process may be carried on for the oversize of all the different screens, and the final undersize, which passes the finest screen, used. The results of these tests may then be tabulated.

"As the result of these tests, it will now be possible to form some idea as to where the values exist, that is, so far as the gold is concerned, but it will be found to be exceedingly difficult to remove the gold from the platinum, except by resorting to the usual refinery methods. The residue remaining after removing the magnetizable product, will consist of other by-products, some of which may have a commercial value, together with a certain amount of fine gold, and just here is where the great field for research is open; for at the present moment, I do not know any process by which these different products may be separated and put into marketable form.

"Care must also be exercised in making a magnetic separation of the magnetizable particles, for if too strong a magnet is used and swept hurriedly through a large

mass of sand, the clusters of iron grains will almost
always pick up and hold in suspension a certain number
of gold particles. For laboratory purposes, I have found
it most convenient to use an ordinary 5 or 6-inch horse-
shoe magnet and to cover the poles with a fine cambric
bag. I then spread out the sand to the thickness of
$\frac{1}{16}$ inch on a piece of paper, and gently pass the covered
magnet back and forth over the surface, and the magnetizable grains will cluster on the outside of the bag.
The covered magnet with its adhering particles is then
placed in a glass bowl and the magnet withdrawn from
the bag and the latter shaken. This operation should be
repeated until all the magnetizable product has been
separated from the sample. The product collected
should again be spread out on a piece of cardboard and
gone over a second time with the magnet; this time,
however, the bag must be drawn tightly over the poles of
the magnet, and the latter must be tapped gently while
picking up the grains, so that any non-magnetizable
gold or platinum particle will be freed from the clusters
and fall back on the cardboard by virtue of its specific
gravity. The use of a bag as described above will be
found to facilitate the operation, and is much quicker
than placing the bare poles of the magnet against the
black sand.

"The gold that has been separated should be carefully
examined under a glass, and notes made regarding its
surface appearance. It will also be a good plan to
make an amalgamation test and find out by weighing
what proportion of the gold will or will not readily
amalgamate.

"As a result of these tests, we now have the following data:

" 1. The weight of fine products resulting per cubic yard of gravel that will have to be treated by concentration, and we can now, therefore, estimate upon the necessary table area per cubic yard of gravel to be washed.

" 2. The weight of black sand concentrated per cubic yard and the approximate gross value per ton of concentrate, and also per cubic yard of gravel.

" 3. The proportion by weight of separation that can be made by magnetic methods.

" 4. The values, if any, that are in direct association with the iron grains and would be lost by this method. Gold and platinum assays should be made.

" 5. The values that remain in the non-magnetic residue subsequent to wind separation of the clean gold, and that with our present knowledge can only be saved by smelting.

" 6. The value of the clean gold that has been caught by the three processes.

" 7. The proportion of gold that will amalgamate, and if the non-amalgamating gold is coarse or fine.

"8. The quantity of residue containing the by-products that need further treatment.

" In the foregoing, I have merely outlined some of the most important tests, though numerous others will no doubt be made before a final solution of the problem can be arrived at. We may, however, now commence to formulate some definite arrangement for a plant to treat the black sand and to indicate the nature of the problems most likely to be encountered. Broadly speaking, all of the coarse material of a placer is value-less, and therefore the sooner it is dumped and gotten rid of, the better. All coarse gold above $\frac{1}{8}$-inch size is easy to catch, and an arrangement of a few riffles in a short line of sluices will take care of that feature. Hence the first and most important step is to size the material, say to $\frac{1}{8}$ inch, and reject the coarse after passing through a sluice. It will be equally necessary to get rid of the excessive quantity of clay and fine product, which might impede the successful operation of further con-centration.

"As dump is an important consideration in many placer mines, it will be necessary to design a plant with as little loss of head room as possible, and I would suggest the following arrangement: At the discharge end of a sluice, where the material is received from mining operations, a grizzly should be placed, having a suffi-cient area to pass all of the fine material and practically all of the water. It should be placed at such an angle that the coarse material will roll off to dump. The grizzly should be provided with taper bars and spaced about $\frac{7}{16}$ inch. All of the fine material will pass by a large sluice to a distributing and hydraulic classifier, which will get rid of the excess water and very light sand and clay held in suspension.

"As a certain amount of float gold might be carried away to the reject, I would suggest that this be passed through some form of amalgamating device, — the Pierce amalgamator being a good machine for this purpose, — and as the light gold has probably a clean surface, it will be fairly easy to amalgamate. It will, however, be advisable to make a number of tests by means of settling vats, to find out whether the value of the gold and by-products saved will be worth the capital outlay necessary to save them.

"The heavy sand from the hydraulic classifiers will pass to some form of classifying sieves, designed to handle a large bulk at a minimum of capital and current expenditure. These should size from $\frac{5}{16}$ to, say, $\frac{1}{8}$ inch, and must be of simple and durable structure, or the placer miner will never bother with them.

"To size all of the material from a placer mine, to reduce the product to be treated to a minimum bulk, economically and without loss, and to deliver it in a form suitable for further concentration and treatment, is a matter that offers a broad field for intelligent design and invention. The quantity and proportion of this sized product to the original gravel will vary according to the nature of the gravel mined; and judging from my experience, it will be least in modern river channels, and greatest in deep, ancient deposits, and may vary from 20 to 60 per cent of the original deposit.

"Some form of concentration must now be adopted to treat the fine material, and although cocoa matting tables with expanded metal riffles have been used fairly successfully for the purpose, they are to be condemned because they are not continuous in operation; and all forms of non-continuous concentration are bad, owing to the fact that the surfaces choke and the values commence to slide over and are lost. This is prevented by frequent clean-ups, but this entails too great an outlay for current expenses, and quickly reduces profits when treating such a low grade product as the sand of placers.

"The Pinder concentrator is a good machine for this purpose. Its capacity is about 40 tons per day, equivalent to handling the product from 60 to 100 cubic yards of gravel. It is also capable of delivering three grades of concentrates and tailing. Any form of concentrator could be adopted, according to the ideas of the mine owner.

"The suggestion is made to the miner to either ship his product to a smelter, or hire a skilled metallurgist. If he desires to be his own metallurgist, proceed as follows: Treat with diluted nitric acid; wash; amalgamate without grinding, to remove the free gold; wash residue through a fine steel-wire sieve. If platinum is present, this can be dissolved by acids, using about 15 times its weight of aqua regia, and precipitate by sal ammoniac. The precipitate of platin-ammonium chloride can be dried and treated by the usual refining methods."

Dr. David Day of the United States Geological Survey had charge of an experimental station at the Lewis and Clark Exposition, the object of which was to concentrate black sands, and extract therefrom the valuable minerals. The method he followed was about as follows: First the sand was dried, and the magnetic minerals separated by magnetic concentration; the sands were then converted into pig iron and steel by the electric furnace.

The tailing contained the gold that was not lost by magnetic concentration, platinum and other non-magnetic sands.

From a commercial standpoint the process was not a success, although it was talked about freely, and some unskilled in mining presumed it was a wonderful discovery of science. To spend $5 to recover $1 is not a scientific discovery, and the whole affair was the cause of much amusement to mining engineers and metallurgists.

On this subject the *Chicago Inter-Ocean* said:

Dr. Day has demonstrated to the people of the Pacific coast that values of untold billions of dollars are to be found in the black sands which line almost the whole Pacific coast and form the banks and river bottoms of almost every river flowing into the Pacific Ocean, from Alaska to Southern California.

That these black sands contain a large percentage of iron has been known for many years. In fact, it has been estimated that, if the iron could be separated from the sand and smelted, the Pacific coast could supply sufficient iron for the markets of the world for thousands of years, but as yet no practical method has been discovered for separating the iron in commercial quantities from the sand.

Machines have been made which will separate the iron from the sand in small quantities when the sand has been thoroughly dried, but the capacity of these machines is so limited, and the cost of drying the sand is so great, as to render them commercially valueless for treating the black sands.

STOREHOUSE OF RARE MINERALS.

But Dr. Day demonstrated at the exposition at Portland that the black sand of the Pacific coast contains many other valuable minerals. Professor Richards, of the Boston School of Technology; Professor Kemp, of Columbia College; J. F. Batchelder, chairman of the mining committee of the Portland Board of Trade; and an able corps of assistants from the various schools of technology, all under the supervision of Dr. Day, carried on an exhaustive analysis of the black sands, from samples taken from thousands of locations along the beaches of the Pacific coast, the beds and bars of rivers flowing into the Pacific Ocean, as well as from the dry beds of ancient rivers in the interior of the Western country.

The preliminary reports of Dr. Day's investigations have already been published by the government at Washington, and the publication of his complete report is eagerly looked for by the people of the Pacific coast.

While it was known that the black sands contained free gold in varying quantities, yet it was considered impossible to recover this gold, on account of its being mixed with the heavy grains of iron, and no practical way of freeing it from the iron being known. However, the government reports of Dr. Day's experiments show that, after he dried the black sand and extracted the iron, by using one of the machines already referred to, it was possible to recover the free gold in paying quantities.

GOLD FROM THE IRON.

But this is not all. The reports show that nearly all the iron in the black sands carries "rusty" gold and that, by separating the iron grains from the sand and treating the iron itself, values in gold were obtained running from $6 to $600 per ton of iron extracted. Furthermore, by placing the sand (from which the heavy iron had been extracted) on the gently oscillating tables, where the separation of the sand into its component parts was made by gravity by pouring water over the incline, it was discovered that the sand contained, besides gold, large quantities of other valuable minerals which were easily separated, such as monazite chromite, garnet, zircon, etc. (some of them worth over $400 per ton, and platinum, which is more valuable even than gold.

As if these discoveries of untold values were not sufficient to set the mining world of the West on fire with excitement and anticipation, Dr. Day, after demonstrating that many of the sands tested contained as high as 600 pounds of iron to the ton, erected a ten-ton electric furnace, and in one short hour, by adding lime and broken coal to the iron, which had been separated from the black sands, smelted the iron into high grade steel, which has stood all tests for purity and toughness, and shown that the black sands of the Pacific coast will stand alongside Norwegian and Swedish iron ore as the mother of steel.

$100,000,000 IN IRON.

The discovery has created even greater interest than the fact that the sands contain gold and platinum and other precious minerals, for the Pacific coast States and Territories always have depended on the East for their supplies of steel and iron. The steel business of the Pacific coast amounts to more than $100,000,000 per year, and on every ton of iron or steel used on the Pacific coast, a cost of $10 per ton has to be added for freight charges, the freight being regulated by the rates by water; for so difficult is it to get sufficient iron to the Pacific coast to supply its rapidly increasing demand for the material, that thousands upon thousands of tons of iron and steel are brought around by Cape Horn by boat from Glasgow, Scotland. This is taking coals to Newcastle with a vengeance, for in almost every foot of sand lapped by the waters of the Pacific Ocean, along the shores of the States of the West, is found the precious iron which the people of the West now go thousands of miles from home to purchase.

The questions of interest that naturally arise are:

1. How did these minerals get into the sand?
2. How extensive are the sand beds?
3. What percentage of iron does the sand contain per ton?
4. How are the iron and minerals to be extracted in commercial quantities?

CONTAINS INCONCEIVABLE WEALTH.

Dr. Day's investigations, for which special purpose the last Congress made a liberal appropriation, show that the black sand is found in enormous, unlimited deposits along the ocean beaches of the Pacific States, but particularly along the coast of Oregon. In some cases the sands of these beaches have been found to contain as high as 40 per cent of iron.

But the greatest values per ton of sand are found in the beds of the rivers flowing into the ocean, for, while they contain less iron (in some cases as low as 10 per cent), still their values run higher in gold and other precious minerals, for these sands are formed by erosion and the breaking down of eruptive rocks which contain minerals and metals of most diverse kind and value in their structure. The dissolution of these mineral rocks along the course of the various rivers for ages past and the erosion caused by these rivers, which have been cutting channels through these rocks for probably hundreds of thousands of years, have separated the rocks into their component parts, forming the black sand, and the rivers, even now, are ceaselessly carrying this precious sand down their entire course and dumping it into the Pacific Ocean, where, by different currents, it is returned to the mainland to build up the black sand beaches of the Pacific coast.

ERODED ROCK RELEASES GOLD.

The pieces of eroded rock from constant friction one upon the other, during their course down the river, gradually grow smaller and smaller as they are moved farther down the river by its currents, gradually dropping their burden of precious minerals, until, on reaching the ocean, the sand contains more iron than anything else, while the higher values in gold have been left farther up the river, where they are found in such large quantities, unmingled with a high percentage of iron, that placer dredging has become an established, lucrative business pursuit.

The black sand has always been known as the "thief" of the placer miner, for sand or gravel that runs higher than 2 per cent in iron cannot be worked profitably by methods now in vogue, for the grains of iron, being nearly as heavy as gold, fill up the riffles and make the separation of gold impossible. In fact, many of the most valuable placer mines are idle to-day on account of the black

sand thief, which the United States Government now tells us contains, in itself, more wealth than human mind can conceive.

Summing up we find

1. That the aggregate amount of magnetic iron contained in
the black sands of the Pacific slope is beyond calculation and practically inexhaustible, as the deposits are being constantly added to
by natural accretion.

2. That, in order to utilize this iron commercially, it must first
be magnetically separated from the sand, without drying, as the
cost of drying a ton of sand for the purpose of recovering 40 to 400
pounds of iron is much greater than the commercial value of the
product warrants.

3. That, with the iron extracted, the high class minerals, such
as gold, platinum, monazite, zircon, etc., which are almost uni-
versally found in these sands, can be easily and cheaply separated
from the surrounding "gangue" by methods of concentration now
in general practical use.

4. That the magnetic iron itself almost always contains sufficient
"rusty" gold to yield a handsome profit by lixiviation, although,
as a rule, not enough to pay for treatment by any other process
now known.

5. That the magnetic iron when separated from the sand can be
reduced in an electric smelter to commercial iron or steel, ready to
supply the Pacific coast markets.

6. That the steel so produced, smelted with cheap electricity
generated from water power, need not exceed $12 per ton, while
the present cost of pig iron on the Pacific coast is over $27 per ton —
a saving of $15 per ton over present prices of pig iron alone.

The government has pointed the way to fabulous fortunes and
gigantic commercial enterprises through the black sands. It now
remains to be seen if scientific discovery and further experiment
will lead the way to the practical, profitable, commercial uses of
the limitless black sand beds of our Western empire.

That this article tells the situation exactly, is evident
from the following copy of a letter written by Dr. Day

himself to the manager of the Inter-Ocean Newspaper Company.

DEPARTMENT OF THE INTERIOR:
UNITED STATES GEOLOGICAL SURVEY.
PORTLAND, ORE., February 21, 1906.

Mr. Samuel S. Sherman, Business Manager, The Inter-Ocean, Chicago, Ill.

DEAR MR. SHERMAN: I thank you very much for your very complimentary article of January 21, which requires but little in order to make it perfect.

I will be glad to give you further reports on the black sand work. Congress has just extended the work, and it will start again in a few days.

The black sand subject is really one of very great interest and is going to aid *very much good citizenship on the Pacific coast,* for it is not a matter of speculation but simply of untiring industry, *with all the personal improvements of character* which come by that kind of work as contrasted with the usual speculation so frequent in the mining industry.   Yours very truly,

DAVID T. DAY.

The only concentrator we know of in the United States that will treat the black sands wet is shown in Fig. 65.

This machine, which is not large, handles dry material readily, to my satisfaction; and Mr. J. F. Batchelder, who was employed by the Government in its investigation of black sands, considers that it preëmpts the whole field opened by the Government's investigations at Portland. The Lovett system consists in raising black sand from a river bottom by means of a pneumatic device or other means.   The material so raised is screened, and the fine material carried to a sluice box in which is placed a Lovett magnetic separator. The tailing is passed to mercury plates. The claim is then made that " this process of

Fig. 65.

recovering free gold is protected by United States patent, which covers the use of any solvent after magnetic separation." The writer ten years ago made experiments on concentrates with cyanide solutions, the concentrates having been extracted from a mass of sand by means of a magnet. We therefore think this claim somewhat broad, and the writer believes that Professor Christy also made experiments of this kind in the University of California.

CHAPTER XIII.

UNITED STATES MINE LAWS.

THE subject of placer mines brings up the question, How can they be obtained? If one has to purchase them, the demand will not be great; if one can locate a claim, the subject becomes interesting to the majority of gold seekers. Information upon this subject, which is well known in the mining States of the West, is entirely unknown in the East, except by those who make a business of mining.

Prior to the Congressional Act of 1866 the ownership of mineral lands was retained by the Government. The agitation for the sale of such lands began in 1850, the object being to make them a source of revenue. The wise policy of leaving such lands open for private development prevailed until 1866, when the uncertainty of titles demanded a change. Possessory rights were all that could be conferred on mining claims, and this could be retained by working and the payment of a small royalty. The law was merely a license to citizens of the United States to go upon mineral lands of the public. The Government owned the land, but placed no claim of ownership on minerals extracted, except so far as license fees or royalty was concerned.

The Act of May 10, 1872, allowed any person a citizen, or one who had declared his intentions to become such, and no others, to locate and hold a mining claim 1500 feet long by 600 feet wide, the claim to be by one person, 1500 linear feet along the course of the mineral vein or lode, subject to location; or any association of persons, severally qualified as above, may make joint location of such claim of 1500 feet; but in no event could a location of a vein or lode, made subsequent to the date mentioned, exceed 1500 feet along the course thereof, whatever should be the number of persons in the company. With regard to the extent of surface ground adjoining a lode or vein, and claimed for the convenient working of the same, it is provided that the lateral extent of location, made after May 10, 1872, shall in no case exceed 300 feet on each side of the middle of the vein at the surface, and that no surface rights shall be limited by any mining regulations to less than 25 feet on each side of the middle of the vein at the surface, except where adverse rights, existing on the 10th of May, 1872, may render such limitations necessary; the end lines of such claims to be in all cases parallel with each other.

Thus it may be seen that no lode claim, located after May 10, 1872, can exceed a parallelogram 1500 by 600 feet, but whether surface ground of that width can be taken depends upon the local or State laws in force in the mining district; but no such laws shall limit a vein or lode claim to less than 1500 feet along its course, nor can surface rights be limited to less than 50 feet in width, unless adverse claims, existing on May 10, 1872, render such lateral limitations necessary. It is provided by the Revised Statutes that miners of each district may make such rules and regulations not in conflict with the laws of the United States, or of the State or Territory in which the districts are situated, governing the location, manner of recording, and amount of work necessary to hold possession of a claim. In order to hold a possessory right to a location made prior to May 10, 1872, not less than $100 worth of labor must be performed or improvements made thereon within one year from the date of such location, and annually thereafter; in default of which the claim will be subject to relocation by any one else having the necessary qualifications, unless the original locator, his heirs, assigns, or legal representatives have resumed work after such failure and before relocation. The expenditures required upon such claims may be made from the surface, or in running a tunnel for their development. The Act of February 11, 1875, provided that where a person or company has run a tunnel for the purpose of developing a lode or lodes the money so expended shall be considered as expended on the said lodes, and the owners shall not be required to perform work on the surface to hold the claim. California has recently passed a new local mining law which in some respects is better than the former law, but in others falls short of what is necessary. The two most needed matters in such State laws are:

What shall constitute a proper marking of a claim so as to avoid litigation? The locator of a claim should therefore not neglect his corner pillars, and make them as conspicuous and durable as possible.

The other matter referred to is, What amount of assessment work shall be done to hold claims, and prevent persons from evading the spirit of the United States statute in regard to assessment work? The locator of a claim should familiarize himself with the local laws of the State or Territory in which he lays out his claim; otherwise it may be "jumped," *i.e.*, have some one take it away from him.

Individual proof of citizenship may be made by affidavit: if a company unincorporated, by the agent's affidavit; if a corporation, by filing a copy of the charter or certificate of incorporation with the secretary of state, county recorder, or with the nearest government land officer — possibly better with each.

Locators against whom no adverse rights rested on the date of the Act of 1872 shall have, on compliance with general and recognized custom, the exclusive right to possession and enjoyment of the surface inclosure and of "all veins, lodes, and ledges which lie under the top or apex of such lines, extended downwards vertically, even though they in their descent extend outside the side lines of such surface locations." (Probably the best expert on the Apex Law is Dr. Rossiter W. Raymond,[1] of New York City. He is one of the framers of the law of 1875, and because of his being at one time at the head of the U. S. Government Survey, he is considered to be the best-informed man on the subject.) The right to such outside parts of veins or ledge is confined to all that lies between "vertical planes drawn downward," as described, so continued that these planes "will intersect the exterior parts of the said veins or ledges." The surface of another claim cannot be entered by the locator or possessor of such lode or vein.

[1] Law of the Apex. R. W. Raymond. A. I. M. E. Transactions.

The land office construes the word *deposit* to be a general term, embracing lodes, ledges, placers, and all other forms in which valuable metals have been discovered. Whatever is recognized as mineral by standard authorities, where the same is found in quality and quantity sufficient to render land sought to be patented more valuable on this account than for the purposes of agriculture, is treated by the land office as coming within the meaning of the act. Lands, therefore, valuable on account of borax, sodium carbonate, nitrate of soda, alum, sulphur, petroleum, and asphalt may be patented.

The first section of the Act of 1872 says, "all valuable mineral deposits." The sixth section uses the term "valuable deposits." This latter section required the Supreme Court to rule petroleum a mineral deposit. This session of Congress, December, 1897, was presented with a bill drafted by Mr. A. H. Ricketts, a mining lawyer of San Francisco, the purpose of which was to recover from railroad companies those lands for which they received patents which lands were known to be mineral before the patents were issued, where they have not passed into the hands of innocent purchasers. Such a bill is eminently proper, and would take away from the railroad companies only lands which they ought never to have received and which the California Miners' Association sought so strenuously to prevent their obtaining.[1] "It is said to be the practice of the railroad companies, when they receive patents for lands to which they know they are not entitled, to transfer them to some outside party who claims to be an innocent purchaser." "The

[1] Mining and Scientific Press, Dec. 18, 1897.

miners generally are determined that the railroad companies shall not hold mining property that never was granted by Act of Congress."

The grant of Congress referred to was, that certain railroads, because of their being built, should have each alternate additional section for ten miles back on each side of the roads as completed, but excludes all minerals except iron and coal from the grant. As fast as the lands were surveyed the companies applied for patents.

Prospectors cannot obtain claims on patented lands, and consequently should keep off them. Mr. Ricketts' proposed law defines the word *mineral* to mean " cinnabar, copper, lead, borax, asphalt, petroleum, oil, salt, and sulphur."

Deposits of fire clay may be patented under the Act of 1872, and so may iron ore deposits be patented as vein or placer claims. Lands more valuable on account of deposits of limestone, marble, kaolin, and mica than for purposes of agriculture may be patented as mineral lands.

The Act further provides that no lode claim can be recorded until after the discovery of the vein or lode within the limits of the ground claimed. The claimant should therefore, prior to recording his claim, unless he can trace the vein on the surface, sink a shaft, run a tunnel or drift to a sufficient depth therein to discover and develop a mineral bearing vein, lode, or crevice; should determine, if possible, the general course of such vein in the direction from the point of discovery, in which direction he will be governed in making the boundary of his claim on the surface; and he should give the course and

direction as nearly as practicable from the discovery
shaft on the claim to some permanent well-known points
of objects, such as, for instance, stone monuments,
blazed trees, the confluence of streams, etc., which may
be in the immediate vicinity, and will serve to perpetuate
and fix the locus of the claim, and render it susceptible of
identification from the description thereon given in the
record of location in the district. He should drive a post
or erect a monument of stones at each corner of his surface ground, and at the point of discovery or discovery
shaft should fix a post, stake, or board, upon which
should be the name given the lode, the name of the locator, the number of feet claimed, and in what direction
from the point of discovery, it being essential that the location notice be filed for record. In addition to the foregoing, the description should state whether the entire
claim of 1500 feet be taken on one side of the point of ✦

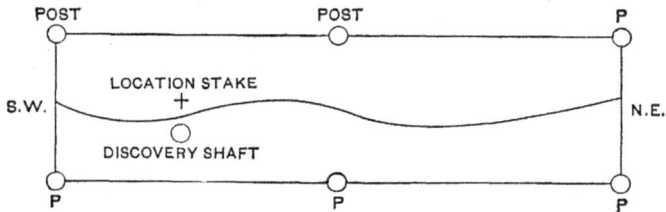

FIG. 66.

discovery or whether it is partly upon the other side, and
in the latter case how many feet are claimed upon each side
of such discovery point.

Parties locating lodes are entitled to all the dips,

spurs, angles, variations, and ledges of the lode coming within the surface ground.

The following diagram will aid the locator in his work (Fig. 66):

MINER'S FORM OF NOTICE.

I, John Doe, hereby give notice that I have this —th day of ———, A.D. 18—, located this, the ———— lode. I claim 1500 feet in and along the vein, linear and horizontal measurement. I claim 1200 feet along the vein running in a northeasterly course from the discovery shaft, and 300 feet running along the vein in a southwesterly course from the discovery shaft. I also claim 150 feet on each side of the vein from center of crevice as surface ground.

JOHN DOE, *Locator*.

In case there are more than two locators, the names of the two should be inserted, and the pronoun "we" where "I" occurs.

There may be intervening claims which will lessen the length or the width of the claim. Within reasonable time after the location shall have been marked on the ground, notice thereof accurately describing the claim in manner aforesaid should be filed for record with the proper recorder of the district, who will thereupon issue the usual certificate of location. District customs are followed in this matter, and should be familiarized by the prospector. These regulations will require that a location certificate be filed with the recorder, in the county in which the lode is situated, within a specified time after its location.

FORM OF RECORDING LOCATION.

STATE OF ———— }
COUNTY OF ———— } ss.:

Know all men by these Presents, That I, John Doe, the undersigned, have this ——th day of ———— A.D., 18—, located and claimed, and by these presents do locate and claim, by right of discovery and location, in compliance with the Mining Acts of Congress, approved May 19th, A.D. 1872, and all subsequent Acts, and with local custom, laws, and regulations, ———— feet linear and horizontal measurement, on the ———— lode, along the vein thereof, with all its dips, angles, and variations, together with ———— feet, running ———— from center of discovery shaft. Said discovery shaft being situated upon said lode, and within the lines of said claim ———— Mining District, County of ————, and State of ————, and further described as follows:

Beginning[1] at the location stake and running in a line southwesterly 300 feet, thence northwesterly to a post 150 feet.

Beginning at this post and running a line northeasterly 1500 feet, to a point marked by post and pile of stones; hence southeasterly 600 feet to a post placed in the ground and marked II; hence southwesterly

---

[1] Explanatory only. See Fig. 66.

1500 feet to a point marked by post and stone pile; and thence 600 feet northwesterly to the point of beginning.

Said lode was located on the —th day of ———, A.D. 18—.

JOHN DOE.

Attest: ——— ———

—th day of ———, A.D. 18—.

In order to hold possessory rights to a claim of 1500 feet of vein or lode located as aforesaid, the Act requires that until a patent shall have been issued therefor not less than $100 worth of labor shall have been expended annually, on the basis adopted by the local mining regulations; in default of which labor or improvements the claim will be subject to relocation by any other party having the necessary qualifications, unless the original locator, his heirs, assigns, or legal representatives have resumed work thereon after such failure and before such relocation.

The importance of attending to these details in the matter of location, labor, and expenditure will be the more readily perceived when it is understood that failure to do so may invalidate the claim. After the patent has been granted, no more assessment work is required.

Five dollars per day is usually allowed for each day of every eight hours' work performed upon a claim for the purpose of holding title or performing the necessary amount of work for the patent, and no other expenses shall be considered as expended for the purpose of holding or protecting title.

PLACER CLAIMS.

The U. S. law prior to May 10, 1872, allowed each person 160 acres or a quarter section of a square mile of placer ground, if located. From the above date all placer claims shall conform as nearly as practicable with the United States system of public surveys, and no such location shall include more than 20 acres for each individual claimant. The provisions of the law are construed by the Commissioner of the General Land Office to mean that after the 9th of July, 1870, no location of placer claim can exceed 160 acres, whatever may be the number of locators associated together, or whatever the local regulation of the district may allow; and that from and after May 10, 1872, no location made by an individual can exceed 20 acres, and no location made by an association of individuals can exceed 160 acres; which location cannot be made by a less number than eight *bona-fide* locators. But whether as much as 20 acres can be located by an individual, or 160 acres by an association, depends entirely upon the mining regulations in force in the respective districts at the date of location; it being held that such mining regulations are in no way enlarged by the statutes, but remain intact in full force with regard to the size of locations, in so far as they do not permit locations in excess of the limits fixed by Congress. A local regulation is valid which provides that a placer claim, for instance, shall not exceed 100 feet square. Congress requires no annual expenditures on placer claims, leaving them subject to the local laws, rules, regulations, and customs of the mining district.

The California Law regarding Placers. — Section 4, Act of 1897, reads:

"The discoverer of placers or other forms of deposit, subject to location and appropriation under mining laws applicable to placers, shall locate his claim in the following manner:

"First. He must immediately post, in a conspicuous place at the point of discovery thereon, a notice or certificate of location thereof, containing:

"*a.* The name of the claim.

"*b.* The name of the locator or locators.

"*c.* The date of discovery and posting of the notice hereinbefore provided for, which shall be considered as the date of location.

"*d.* A description of the claim by reference to legal subdivisions or sections, if the location is made in conformity with the public surveys; otherwise, a description with reference to some natural object or permanent monument as will identify the claim; and where such claim is located by legal subdivisions of the public surveys such location shall, notwithstanding that fact, be marked by the locator upon the ground, the same as other locations.

"Second. Within thirty days from the date of such discovery he must record such notice or certificate of location in the office of the county recorder of the county in which such discovery is made, and so distinctly mark his location on the ground that its boundaries can be readily traced.

"Third. Within sixty days from the date of the discovery the discoverer shall perform labor upon such location or claim in developing same to an amount which shall be equivalent in the aggregate to at least ten dollars ($10) worth of such labor for each twenty acres, or fractional part thereof, contained in such location or claim.

"Fourth. A failure to perform such labor within said time shall cause all rights under such location to be forfeited, and the discovery thereby shall at once be open to location by qualified locators other than the preceding locators, but shall not in any event be open to location by such preceding locators, and any labor performed by them thereon shall not inure to the benefit of any subsequent locator thereof.

"Fifth. Such locator shall, upon the performance of such labor, file with the recorder of the county an affidavit showing such performance, and generally the nature and kind of work so done."

Section 5 of the same Act reads: "The affidavit provided for in the last section, and the aforesaid placer notice or certificate of location when filed for location, shall be deemed and considered as *prima facie* evidence of the facts therein recited. A copy of such certificate, notice, or affidavit, certified by the county recorder, shall be admitted in evidence in all actions or proceedings with the same effect as the original."

In locating a claim, if the above directions are closely followed, no matter what the locality, the prospector will generally have complied with the law. However, it is better to have the local laws well understood whenever possible.

The United States statutes provide "water rights."

1. That as a condition of sale, in the absence of legis-

lation by Congress, the legislature of a State or Territory may provide rules for working mines, involving easements, drainage, and other necessary conditions; these to be expressed in the patent.

2. All prior rights, arising from possession, in the use of water, and recognized by local laws, etc., or judicial decisions, shall be regarded as vested, and shall be protected. This right of way is also granted and confirmed. Damages are to accrue if a land settler's rights are interfered with.

3. All land patents shall be subject to vested and accrued water rights, including ditches and reservoirs. Officers of the U. S. Land Office are required to file with the General Land Office the local laws on such matters. Water privileges are, since the Act of May 10, 1872, located in the same manner as mines, subject to local regulations, *i.e.*, by definitely locating the five acres by monuments, and recording with the district or county recorder. If the local rules and decisions of courts make the privilege forfeitable for non-use, another party may come in and claim the water right. The Federal courts have decided that the right of way to construct flumes or ditches over public lands is unquestionable. It has also been decided that the miner's right to water, within "reasonable limits," is not to be questioned. "It must be exercised, however, with due regard to the general condition and needs of the community, and cannot vest as an individual monopoly."

MILL SITES.

Land, non-mineral in character, and not contiguous to the vein or lode, used by the locator and proprietor for mining or milling purposes, can be included in any application for patent, to an extent not to exceed five acres, and subject to examination and payment as fixed for the superficies of the lode. The owner of a quartz mill or reduction mill, not a mine owner in connection therewith, may also receive a mill site patent. Such sites are located under the mining act, and in compliance with local law and customs as recognized. Such possessory rights give title also to all growing timber thereon. There must in every case be given satisfactory proof of the non-mineral character of the site, and the improvements thereon must be equal to $500. A mill passes to a railroad if located on railroad land grant, and presumably to some-one else if located on another's ground. The location of a mill site is of considerable importance, and should be examined thoroughly. It must be surveyed by a Deputy Mineral Land surveyor, and recorded the same as a lode or placer claim.

CHAPTER XIV.

THE MINING REGULATIONS FOR THE CANADIAN YUKON.

WE give below, substantially in full, the new regulations governing placer mining and dredging in the provisional district of the Yukon, as approved by Order in Council dated Ottawa, January 18, 1898. These regulations constitute the mining law under which all operations must be conducted in that portion of the Yukon region which is in Canadian territory; and the Dominion Government is making provisions for their strict enforcement. The regulations are as follows:

INTERPRETATION.

"Free Miner" shall mean a male or female over the age of 18, but not under that age, or joint-stock company, named in, and lawfully possessed of, a valid existing free miner's certificate, and no other.

"Legal Post" shall mean a stake standing not less than 4 feet above the ground and flatted on two sides for at least 1 foot from the top. Both sides so flatted shall measure at least 4 inches across the face. It shall also mean any stump or tree cut off and flatted or faced to the above height and size.

"Close Season" shall mean the period of the year during which placer mining is generally suspended.

The period to be fixed by the mining recorder in whose district the claim is situated.

"Mineral" shall include all minerals whatsoever other than coal.

"Joint-stock Company" shall mean any company incorporated for mining purposes under a Canadian charter or licensed by the Government of Canada.

"Mining Recorder" shall mean the official appointed by the gold commissioner to record applications and grant entries for claims in the mining divisions into which the commissioner may divide the Yukon District.

Free Miners and Their Privileges.

1. Every person over but not under 18 years of age, and every joint-stock company, shall be entitled to all the rights and privileges of a free miner, under these regulations and under the regulations governing quartz mining, and shall be considered a free miner upon taking out a free-miner's certificate. A free miner's certificate issued to a joint-stock company shall be issued in its corporate name. A free-miner's certificate shall not be transferable.

2. A free-miner's certificate may be granted for one year to run from the date thereof or from the expiration of the applicant's then existing certificate, upon the payment therefor of the sum of $10, unless the certificate is to be issued in favor of a joint-stock company, in which case the fee shall be $50 for a company having a nominal capital of $100,000 or less, and for a company having a nominal capital exceeding $100,000,

the fee shall be $100. Only one person or joint-stock company shall be named in a certificate.

3. Gives form of miner's certificate, and adds: This certificate shall also grant to the holder thereof the privileges of fishing and shooting, subject to the provisions of any act which has been passed, or which may hereafter be passed, for the protection of game and fish; also the privilege of cutting timber for actual necessities, for building houses, boats, and for general mining operations; such timber, however, to be for the exclusive use of the miner himself, but such permission shall not extend to timber which may have been heretofore or which may hereafter be granted to other persons or corporations.

4. Free-miner's certificates may be obtained by applicants in person at the Department of the Interior, Ottawa, or from the agents of Dominion Lands at Winnipeg, Manitoba; Calgary, Edmonton, Prince Albert, in the Northwest Territories; Kamloops and New Westminster, in the Province of British Columbia; at Dawson City in the Yukon District; also from agents of the government at Vancouver and Victoria, British Columbia, and at other places which may from time to time be named by the Minister of the Interior.

5. If any person or joint-stock company shall apply for a free-miner's certificate at the agent's office during his absence, and shall leave the fee required by these regulations, with the officer or other person in charge of said office, he or it shall be entitled to have such certificate from the date of such application; and any free miner shall at any time be entitled to obtain

a free-miner's certificate commencing to run from the expiration of his then existing free-miner's certificate, provided that when he applies for such certificate he shall produce to the agent, or in case of his absence shall leave with the officer or other person in charge of the agent's office, such existing certificate.

6. If any free-miner's certificate be accidentally destroyed or lost, the owner thereof may, on payment of a fee of $2, have a true copy of it, signed by the agent, or other person by whom or out of whose office the original was issued. Every such copy shall be marked "Substituted Certificate"; and unless some material irregularity be shown in respect thereof, every original or substituted free-miner's certificate shall be evidence of all matters therein contained.

7. No person or joint-stock company will be recognized as having any right or interest in or to any placer claim, quartz claim, mining lease, bed-rock flume grant, or any minerals in any ground comprised therein, or in or to any water right, mining ditch, drain, tunnel, or flume, unless he or it and every person in his or its employment shall have a free-miner's certificate unexpired. And on the expiration of a free-miner's certificate the owner thereof shall absolutely forfeit all his rights and interest in or to any placer claim, mining lease, bed-rock flume grant, and any minerals in any ground comprised therein, and in or to any and every water right, mining ditch, drain, tunnel, or flume, which may be held or claimed by such owner of such expired free-miner's certificate, unless such owner shall, on or before the day following the expiration of such certifi-

cate, obtain a new free-miner's certificate. Provided, nevertheless, that should any co-owner fail to keep up his free-miner's certificate such failure shall not cause a forfeiture or act as an abandonment of the claim, but the interest of the co-owner who shall fail to keep up his free-miner's certificate shall, *ipso facto*, be and become vested in his co-owners, *pro rata* according to their former interests; provided, nevertheless, that a shareholder in a joint-stock company need not be a free miner, and, though not a free miner, shall be entitled to buy, sell, hold or dispose of any shares therein.

8. Every free miner shall, during the continuance of his certificate, but not longer, have the right to enter, locate, prospect, and mine for gold and other minerals upon any lands in the Yukon District, whether vested in the Crown or otherwise, except upon government reservations for town sites, land which is occupied by any building, and any land falling within the curtilage of any dwelling-house, and any land lawfully occupied for placer-mining purposes, and also Indian reservations.

9. Previous to any entry being made upon lands lawfully occupied, such free miner shall give adequate security, to the satisfaction of the mining recorder, for any loss or damage which may be caused by such entry; and after such entry he shall make full compensation to the occupant or owner of such lands for any loss or damage which may be caused by reason of such entry; such compensation, in case of dispute, to be determined by a court having jurisdiction in mining disputes, with or without a jury.

NATURE AND SIZE OF CLAIMS.

10. A creek or gulch claim shall be 250 feet long measured in the general direction of the creek or gulch. The boundaries of the claim which run in the general direction of the creek or gulch shall be lines along bed or rim rock 3 feet higher than the rim or edge of the creek, or the lowest general level of the gulch within the claim, so drawn or marked as to be at every point

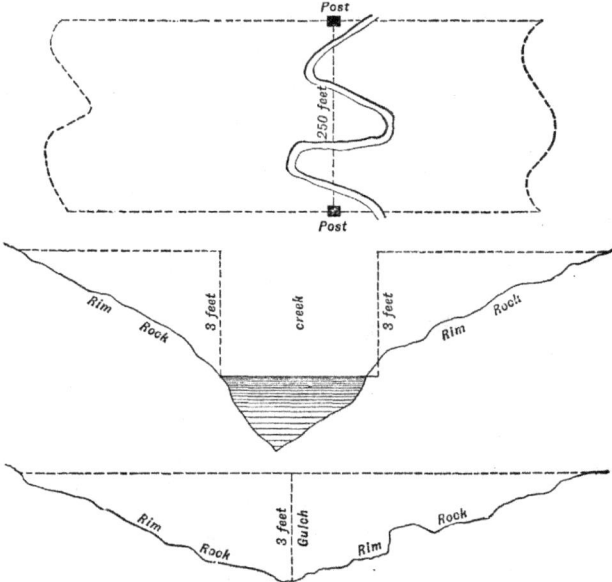

NO. 1.—PLAN AND SECTIONS OF CREEK AND GULCH CLAIMS.

3 feet above the rim or edge of the creek or the lowest general level of the gulch, opposite to it at right angles to the general direction of the claim for its length, but such boundaries shall not in any case exceed 1000 feet

on each side of the center of the stream or gulch. (See Diagram No. 1.)

11. If the boundaries be less than 100 feet apart horizontally, they shall be lines traced along bed or rim rock 100 feet apart horizontally, following as nearly as practicable the direction of the valley for the length of the valley for the length of the claim. (See Diagram No. 2.)

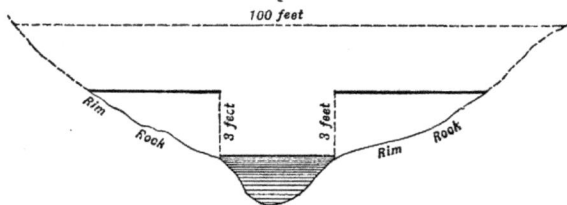

No. 2.—Side Boundaries less than 100 Ft. Apart.

12. A river claim shall be situated only on one side of the river and shall not exceed 250 feet in length, measured in the general direction of the river. The other boundary of the claim which runs in the general direction of the river shall be lines along bed or rim rock 3 feet higher than the rim or edge of the river within the claim so drawn or marked as to be at every point 3 feet above the rim or edge of the river opposite to it at right angles to the general direction of the claim for its length, but such boundaries shall not in any case be less than 250 feet or exceed a distance of 1000 feet from low-water mark of the river. (See Diagram No. 3.)

13. A "hill claim" shall not exceed 250 feet in length, drawn parallel to the main direction of the stream or ravine on which it fronts. Parallel lines drawn from

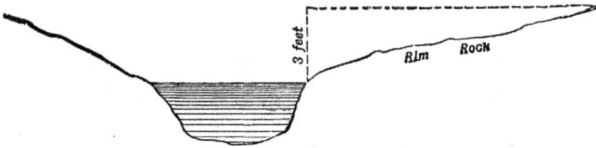

No. 3.—Section of River Claim.

each end of the base at right angles thereto, and running to the summit of the hill (provided the distance does not exceed 1000 feet), shall constitute the end boundaries of the claim.

14. All other placer claims shall be 250 feet square.

15. Every placer claim shall be as nearly as possible rectangular in form, and marked by two legal posts firmly fixed in the ground in the manner shown in Diagram No. 4. The line between the two posts shall

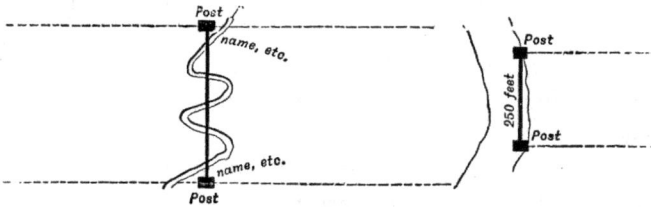

No. 4.—Staking Creek and River Claims.

be well cut out so that one post may, if the nature of the surface will permit, be seen from the other. The flatted side of each post shall face the claim, and on each post shall be written on the side facing the claim, a legible note stating the name or number of the claim, or both if possible, its length in feet, the date when staked, and the full Christian and surname of the locator.

16. Every alternate 10 claims shall be reserved for the Government of Canada. That is to say, when a claim is located the discoverer's claim and 9 additional claims adjoining each other and numbered consecutively will be open for registration. Then the next 10 claims of 250 feet each will be reserved for the Government, and so on. The alternate group of claims reserved for the Crown shall be disposed of in such manner as may be decided by the Minister of the Interior.

17. The penalty for trespassing upon a claim reserved for the Crown shall be immediate cancellation by the mining recorder for any entry or entries which the person trespassing may have obtained, whether by original entry or purchase, for a mining claim, and the refusal by the mining recorder of the acceptance of any application which the person trespassing may at any time make for a claim. In addition to such penalty, the mounted police, upon a requisition from the mining recorder to that effect, shall take the necessary steps to eject the trespasser.

18. In defining the size of claims, they shall be measured horizontally irrespective of inequalities on the surface of the ground.

19. If any free miner or party of free miners discover a new mine, and such discovery shall be estab-

lished to the satisfaction of the mining recorder, creek, river, or hill, claims of the following size shall be allowed, namely: To one discoverer, one claim, 500 feet in length. To a party of two discoverers, two claims, amounting together to 1000 feet in length. To each member of a party beyond two in number, a claim of the ordinary size only.

20. A new stratum of auriferous earth or gravel situated in a locality where the claims have been abandoned shall for this purpose be deemed a new mine, although the same locality shall have been previously worked at a different level.

21. The forms of application for a grant for placer mining, and the grant of the same, shall be those contained in forms H and I in the schedule hereto.

22. A claim shall be recorded with the mining recorder in whose district it is situated, within 10 days after the location thereof, if it is located within 10 miles of the mining recorder's office. One extra day shall be allowed for every additional 10 miles or fraction thereof.

23. In the event of the claim being more than 100 miles from a recorder's office, and situated where other claims are being located, the free miners, not less than five in number, are authorized to meet and appoint one of their number a " Free-miners' Recorder," who shall act in that capacity until a mining recorder is appointed by the gold commissioner.

24. The free-miners' recorder shall, at the earliest possible date after his appointment, notify the nearest Government mining recorder thereof, and upon the arrival of the Government mining recorder he shall

deliver to him his records and the fees received for recording the claims. The Government mining recorder shall then grant to each free miner whose name appears in the records an entry for his claim on form I of these regulations, provided an application has been made by him in accordance with form H thereof. The entry to date from the time the free-miners' recorder recorded the application.

25. If the free-miners' recorder fails within three months to notify the nearest Government mining recorder of his appointment, the claims which he may have recorded will be cancelled.

26. During the absence of the mining recorder from his office, the entry for a claim may be granted by any person whom he may appoint to perform his duties in his absence.

27. Entry shall not be granted for a claim which has not been staked by the applicant in person in the manner specified in these regulations. An affidavit that the claim was staked out by the applicant shall be embodied in form H in the schedule hereto.

28. An entry fee of $15 shall be charged the first year, and an annual fee of $15 for each of the following years. This provision shall apply to claims for which entries have already been granted.

29. A statement of the entries granted and fees collected shall be rendered by the mining recorder to the gold commissioner at least every three months, which shall be accompanied by the amount collected.

30. A royalty of 10 per cent on the gold mined shall be levied and collected on the gross output of each

claim. The royalty may be paid at banking offices to be established under the auspices of the Government of Canada, or to the gold commissioner, or to any mining recorder authorized by him. The sum of $2500 shall be deducted from the gross annual output of a claim when estimating the amount upon which royalty is to be calculated, but this exemption shall not be allowed unless the royalty is paid at a banking office or to the gold commissioner or mining recorder. When the royalty is paid monthly or at longer periods, the deduction shall be made ratable on the basis of $2500 per annum for the claim. If not paid to the bank, gold commissioner, or mining recorder, it shall be collected by the custom officials or police officers when the miner passes the posts established at the boundary of a district. Such royalty to form part of the consolidated revenue, and to be accounted for by the officers who collect the same in due course. The time and manner in which such royalty shall be collected shall be provided for by regulations to be made by the gold commissioner.

31. Default in payment of such royalty, if continued for 10 days after notice has been posted on the claim in respect of which it is demanded, or in the vicinity of such claim, by the gold commissioner or his agent, shall be followed by cancellation of the claim. Any attempt to defraud the Crown by withholding any part of the revenue thus provided for, by making false statements of the amount taken out, shall be punished by cancellation of the claim in respect of which fraud or false statements have been committed or made. In respect to the facts as to such fraud or false statements

or non-payment of royalty, the decision of the gold commissioner shall be final.

32. After the recording of a claim the removal of any post by the holder thereof or by any person acting in his behalf, for the purpose of changing the boundaries of his claim, shall act as a forfeiture of the claim.

33. The entry of every holder of a grant for placer mining must be renewed and his receipt relinquished and replaced every year, the entry fee being paid each time.

34. The holder of a creek, gulch, or river claim may, within 60 days after staking out the claim, obtain an entry for a hill claim adjoining it, by paying to the mining recorder the sum of $100. This permission shall also be given to the holder of a creek, gulch, or river claim obtained under former regulations, provided that the hill claim is available at the time an application is made therefor.

35. No miner shall receive a grant of more than one mining claim in a mining district, the boundaries of which shall be defined by the mining recorder, but the same miner may also hold a hill claim, acquired by him under these regulations in connection with a creek, gulch, or river claim, and any number of claims by purchase; and any number of miners may unite to work their claims in common, upon such terms as they may arrange, provided such agreement is registered with the mining recorder and a fee of $5 paid for each registration.

36. Any free miner or miners may sell, mortgage, or dispose of his or their claims, provided such dis-

posal be registered with, and a fee of $2 paid to, the mining recorder, who shall thereupon give the assignee a certificate in the form J in the schedule hereto.

37. Every free miner shall during the continuance of his grant have the exclusive right of entry upon his own claim for the mine-like working thereof, and the construction of a residence thereon, and shall be entitled exclusively to all the proceeds realized therefrom, upon which, however, the royalty prescribed by these regulations shall be payable; provided that the mining recorder may grant to the holders of other claims such right of entry thereon as may be absolutely necessary for the working of their claims, upon such terms as may to him seem reasonable. He may also grant permits to miners to cut timber thereon for their own use.

38. Every free miner shall be entitled to the use of so much of the water naturally flowing through or past his claim, and not already lawfully appropriated, as shall, in the opinion of the mining recorder, be necessary for the due working thereof, and shall be entitled to drain his own claim free of charge.

39. A claim shall be deemed to be abandoned and open to occupation and entry by any person when the same shall have remained unworked on working days, excepting during the close season, by the grantee thereof or by some person on his behalf for the space of 72 hours, unless sickness or other reasonable cause be shown to the satisfaction of the mining recorder, or unless the grantee is absent on leave given by the mining recorder, and the mining recorder, upon obtaining evidence satisfactory to himself that this provision is not

being complied with, may cancel the entry given for a claim.

40. If any cases arise for which no provision is made in these regulations, the provisions of the regulations governing the disposal of mineral lands other than coal lands, approved by His Excellency the Governor in Council on November 9, 1889, or such other regulations as may be substituted therefor, shall apply. (Appended to Section 40 are the forms for applications, certificates, etc., referred to in the text.)

REGULATIONS GOVERNING RIVER–BED DREDGING FOR GOLD.

The following are the regulations for the issues of leases to persons or companies who have obtained a free-miner's certificate in accordance with the provisions of the regulations governing placer mining in the Provisional District of Yukon, to dredge for minerals other than coal in the submerged beds or bars of rivers in the Provisional District of Yukon, in the Northwest Territories:

1. The lessee shall be given the exclusive right to subaqueous mining and dredging for all minerals with the exception of coal in and along an unbroken extent of five miles of a river following its sinuosities, to be measured down the middle thereof, and to be described by the lessee in such manner as to be easily traced on the ground; and although the lessee may also obtain as many as five other leases, each for an unbroken extent of five miles of a river, so measured and described, no more than six such leases will be issued in favor of an

individual or company, so that the maximum extent of river in and along which any individual or company shall be given the exclusive right above mentioned, shall under no circumstances exceed 30 miles. The lease shall provide for the survey of the leasehold under instructions from the Surveyor General, and for the filing of the returns of survey in the Department of the Interior within one year from the date of the lease.

2. The lease shall be for a term of 20 years, at the end of which time all rights vested in, or which may be claimed by the lessee under his lease, are to cease and determine. The lease may be renewable, however, from time to time thereafter in the discretion of the Minister of the Interior.

3. The lessee's right of mining and dredging shall be confined to the submerged beds or bars in the river below low-water mark, that boundary to be fixed by its position on the first day of August in the year of the date of the lease.

4. The lease shall be subject to the rights of all persons who have received or who may receive entries for claims under the Placer-mining Regulations.

5. The lessee shall have at least one dredge in operation upon the five miles of river leased to him, within two seasons from the date of his lease, and if, during one season when operations can be carried on, he fails to efficiently work the same to the satisfaction of the Minister of the Interior, the lease shall become null and void unless the Minister of the Interior shall otherwise decide. Provided that when any company or individual has obtained more than one lease, one

dredge for each 15 miles or portion thereof shall be held to be compliance with this regulation.

6. The lessee shall pay a rental of $100 per annum for each mile of river so leased to him. The lessee shall also pay to the Crown a royalty of 10 per cent. on the output in excess of $15,000, as shown by sworn returns to be furnished monthly by the lessee to the gold commissioner during the period that dredging operations are being carried on; such royalty, if any, to be paid with each return.

7. The lessee who is the holder of more than one lease shall be entitled to the exemption as to royalty provided for by the next preceding regulation to the extent of $15,000 for each five miles of river for which he is the holder of a lease; but the lessee under one lease shall not be entitled to the exemption as to royalty provided by the next two preceding regulations, where the dredge or dredges used by him have been used in dredging by another lessee, or in any case in respect of more than 30 miles.

8. The lessee shall be permitted to cut free of all dues, on any land belonging to the Crown, such timber as may be necessary for the purposes of his lease, but such permission shall not extend to timber which may have been heretofore or may hereafter be granted to other persons or corporations.

9. The lessee shall not interfere in any way with the general right of the public to use the river in which he may be permitted to dredge, for navigation and other purposes; the free navigation of the river shall not be impeded by the deposit of tailings in such manner as to

form bars or banks in the channel thereof, and the current or stream shall not be obstructed in any material degree by the accumulation of such deposits.

10. The lease shall provide that any person who has received or who may receive entry under the Placer-mining Regulations shall be entitled to run tailings into the river at any point thereon, and to construct all works which may be necessary for properly operating and working his claim. Provided that it shall not be lawful for such person to construct a wing dam within 1000 feet from the place where any dredge is being operated, nor to obstruct or interfere in any way with the operation of any dredge.

11. The lease shall reserve all roads, ways, bridges, drains, and other public works, and all improvement now existing, or which may hereafter be made in, upon, or under any part of the river, and the power to enter and construct the same, and shall provide that the lessee shall not damage nor obstruct any public ways, drains, bridges, works, and improvements now or hereafter to be made upon, in, over, through, or under the river; and that he will substantially bridge or cover and protect all the cuts, flumes, ditches, and sluices, and all pits and dangerous places at all points where they may be crossed by a public highway or frequented path or trail, to the satisfaction of the Minister of the Interior.

12. That the lessee, his executors, administrators, or assigns, shall not nor will assign, transfer, or sublet the demised premises, or any part thereof, without the consent in writing of the Minister first had and obtained.

UNITED STATES

LAW RELATIVE TO RIVER DREDGING.

Rivers or streams in a defined channel belong to the State. To dredge river beds requires either a grant or prescription from the State, in the absence of any definite legislation regulating this industry. There is scarcely any doubt that the State would grant permission to mine any river bed within her borders provided navigation were not hindered or obstructed or riparian rights interfered with, but such sanction should be obtained prior to commencing work. In the case of a non-navigable stream flowing within the borders of one's land, the land under the water belongs to the landowner; or the water, to the State. In such a case the right to dredge is unquestionable. Or in case of two landowners adjoining on opposite sides either one may dredge his half and be convicted of trespass if he oversteps the boundary. The owner or owners must not overstep the mark and injure land or watercourse below their property, otherwise they may be enjoined.

The dredging company may purchase a piece of land and work their dredge along the river bank. They must not, however, change the course of the stream or divert it from the riparian owner opposite, although they may work as far inland on their own property as they desire, and have the usufruct of the stream. The chances are that dredgers if they pollute the streams or make them muddy will have trouble with riparian owners in the States for creating a nuisance. The washing of ore and discoloring the water of the New River, Virginia, has caused much comment, and an attempt has been made in Congress to suppress it.

The Anthracite Mine Operators are compelled to impound the material resulting from coal washing, where once they permitted it to run into the streams and rivers.

If the stream belongs to the public domain, twenty acres can be located, or a dredger may work up and down the stream (provided it does not work on a located claim) without interference.

Navigable rivers are under the supervision of the United States ; other streams belong to the States through which they flow. Mineral lands under rivers belong to the State, and can be obtained from the State by proper legal proceedings.

CHAPTER XV.

THE following tables have been computed by data obtained from careful experiments made by the ablest engineers.

They will therefore assist the unskilled as well as the skilled in many problems. However, to thoroughly understand the subject, one should purchase a text-book on hydraulics.

These tables are reliable, and will prove correct as far as they go.

The whole subject has been touched upon in the preceding pages, so that any one who has carefully read them should understand the tables at a glance, and be able to apply them in practice.

EXPLANATION OF TABLE.

The table furnishes an exceedingly simple method of determining the value of *free gold* in a ton of gold-bearing quartz, or a cubic yard of auriferous gravel.

Take a sample of four (4) pounds of quartz, pulverize it to the usual fineness for horning, wash it carefully by batea or other means, amalgamate the gold by the application of quicksilver, volatilize the quicksilver by blow-pipe or otherwise, weigh the resulting button, and the value given in the table opposite such weight will be the value in free gold per ton of 2000 pounds of quartz.

Example. — Sample of four pounds produces button weighing one grain, the fineness of the gold being 830; then the value of one ton of such quartz will be $17.87.

If the sample of 4 pounds should produce a button weighing say two and four-tenths ($2\frac{4}{10}$) grains, then the value of such quartz would be (875 fine) as follows, viz.:

Opposite 2 grains, 875 fine, value $37.68
Opposite $\frac{4}{10}$ grains, 875 fine, value   7.53
    Total value per ton (2000 lbs.) . . $45.21

GOLD TABLE

FOR DETERMINING THE VALUE OF FREE GOLD PER TON (2000 LBS.)
OF QUARTZ OR CUBIC YARD OF GRAVEL,

Prepared by

MELVILLE ATWOOD, Esq., F.G.S., *Consulting Mining Engineer.*

Weight Washed Gold. 4-lb Sample. Grains.	Fineness, 780. Value per Oz. $16.12.	Fineness, 830. Value per Oz. $17.15.	Fineness, 875. Value per Oz. $18.08.	Fineness, 920. Value per Oz. 19.01.
5 grains	$83.97	$89.36	$94.20	$99.05
4 "	67.18	71.49	75.36	79.24
3 "	50.38	53.61	56.52	59.43
2 "	33.59	35.74	37.68	39.62
1 "	16.79	17.87	18.84	19.81
.9 "	15.11	16.08	16.95	17.82
.8 "	13.43	14.29	15.07	15.84
.7 "	11.75	12.51	13.19	13.86
.6 "	10.07	10.73	11.30	11.88
.5 "	8.40	8.93	9.42	9.90
.4 "	6.71	7.14	7.53	7.92
.3 "	5.03	5.36	5.65	5.94
.2 "	3.36	3.57	3.76	3.96
.1 "	1.68	1.78	1.88	1.98

GOLD VALUE OF A CUBIC YARD OF GRAVEL.

To determine the gold value of a cubic yard of auriferous gravel the foregoing table can be used.

Take a sample of sixty (60) pounds of gravel, pulverize it, and carefully wash it by batea, pan, or otherwise; amalgamate the gold, volatilize the quicksilver, weigh the button, and in column in foregoing table, opposite the weight, will be found the gold value of a cubic yard of the gravel.

Example. — Sample of sixty pounds produces button weighing one grain, the fineness of the gold being 780; then the value of one cubic yard of such gravel would be $1.67. This is arrived at by pointing off one point, or dividing the value given in table by 10.

If the sample of sixty pounds yields a button weighing 1 grain and two-tenths ($1\frac{2}{10}$ grains), then the value of the gravel per cubic yard would be — gold being 920 fine — as follows:

Opposite 1 grain,   920 fine,   value $1.98

Opposite $\frac{2}{10}$ grain,   920 fine,   value   .40 +

Total value cubic yard . . . . $2.38 +

This table is prepared upon the following basis of weights, viz.: A sample of 4 pounds of quartz is the one-five-hundredth part in weight of a ton of 2000 pounds, and the gold values given are reduced to this proportion.

Eighteen cubic feet of gravel in bank will weigh one ton, or 2000 pounds, and a cubic yard, or 27 cubic feet, will weigh 3000 pounds, or $1\frac{1}{2}$ tons; and 60 pounds being the one-fiftieth part of the weight of a cubic yard, then the relative proportion of the weight of quartz to gravel is as 50 to 500, or 1 to 10.

HYDRAULICS.

1 gallon of water = 231 cubic inches and weighs 8.3389 pounds, figured at 8⅓ pounds.

1 cubic foot of water = 1728 cubic inches and weighs 62.3793 pounds, figured at 62.5.
contains 7.48052 gallons, usually figured at 7.5.

A column of water 2.31 feet high gives 1 lb. pressure on each square inch of its base.

A column of water 1 ft. high will give a pressure of .434 lb. on each square inch of base. Usually reckoned at 5 lbs. per ft. in height.

Doubling the diameter of a pipe increases its area four times, hence its capacity.

Doubling the diameter of a pipe increases its frictional rubbing-surface two times.

To double the quantity of water flowing through a pipe under a given head requires eight times the power.

27 154 inches of water will spread 1 inch deep over 1 acre of ground, and weigh 101 tons.

A foot-pound of work is the expenditure of power required to raise one pound one foot high in one minute.

A horse-power is 33,000 foot-pounds, or what a strong horse can do 10 hours daily every minute in the day. Average horses can do but 22,000 ft.-lbs per minute.

To find the horse-power required to raise water: Multiply the number of pounds of water to be raised per minute by the height from the level of the water to the level of discharge.

TABLES FOR CALCULATING THE HORSE-POWER OF WATER.

MINERS'-INCH TABLE.

The following table gives the horse-power of one miners' inch of water under heads from one up to eleven hundred feet. This inch equals 1½ cubic feet per minute.

Head in Feet.	Horse-power.	Head in Feet.	Horse-power.
1	.0024147	320	.772704
20	.0482294	330	.796851
30	.072441	340	.820998
40	.096588	350	.845145
50	.120735	360	.869292
60	.144882	370	.893439
70	.169029	380	.917586
80	.193176	390	.941733
90	.217323	400	.965880
100	.241470	410	.990027
110	.265617	420	1.014174
120	.289764	430	1.038321
130	.313911	440	1.062468
140	.338058	450	1.086615
150	.362205	460	1.110762
160	.386352	470	1.134909
170	.410499	480	1.159056
180	.434646	490	1.183206
190	.458793	500	1.207350
200	.482940	520	1.255644
210	.507087	540	1.303938
220	.531234	560	1.352232
230	.555381	580	1.400526
240	.579528	600	1.448820
250	.603675	650	1.569555
260	.627822	700	1.690290
270	.651969	750	1.811025
280	.676116	800	1.931760
290	.700263	900	2.173230
300	.724410	1000	2.414700
310	.748557	1100	2.656170

WHEN THE EXACT HEAD IS FOUND IN ABOVE TABLE.

Example.—Have 100-foot head and 50 inches of water. How many horse-power?

By reference to above table the horse-power of 1 inch under 100 feet head is .241470. This amount multiplied by the number of inches, 50, will give 12.07 horse-power.

CUBIC-FEET TABLE.

The following table gives the horse-power of one cubic foot of water per minute under heads from one up to eleven hundred feet:

Head in Feet.	Horse-power.	Head in Feet.	Horse-power.
1	.0016098	320	.515136
20	.032196	330	.531234
30	.048294	340	.547332
40	.064392	350	.563430
50	.080490	360	.579528
60	.096588	370	.595626
70	.112686	380	.611724
80	.128784	390	.627822
90	.144892	400	.643920
100	.160980	410	.660018
110	.177078	420	.676116
120	.193176	430	.692214
130	.209274	440	.708312
140	.225372	450	.724410
150	.241470	460	.740508
160	.257568	470	.756606
170	.273666	480	.772704
180	.289764	490	.788802
190	.305862	500	.804900
200	.321960	520	.837096
210	.338058	540	.869292
220	.354156	560	.901488
230	.370254	580	.933684
240	.386352	600	.965880
250	.402450	650	1.046370
260	.418548	700	1.126860
270	.434646	750	1.207350
280	.450744	800	1.287840
290	.466842	900	1.448820
300	.482940	1000	1.609800
310	.499038	1100	1.770780

WHEN EXACT HEAD IS NOT FOUND IN TABLE.

Take the horse-power of 1 inch under 1-foot head and multiply by the number of inches, and then by number of feet head. The product will be the required horse-power.

Note.—The above formula will answer for the cubic-feet table, by substituting the equivalents therein for those of miners' inches.

Horse-power given in above table equal 85 per cent of theoretical power.

FLOW OF WATER THROUGH CLEAN IRON PIPES.

Remarks. — In the analysis of the flow of water, the total head is divided into three parts,: viz., 1st, that portion of the head due to the velocity; 2d, that portion which overcomes the resistance of entry; and 3d, that portion which overcomes the resistance within the pipe. In long pipes, the two former portions as compared with the latter portion of the total head are quite small. In this table the greatest velocity in any pipe is 13.445 feet per second, due to 4.2 feet, the sum of the first and second portions of the total head, while the third portion of the head is 211.2 feet. The head or fall in this table refers to the third portion of the total head. This table has been computed on the assumption that the length of any pipe is not less than 1000 times its diameter.

Question: The fall being 52.8 feet per mile, what will be the flow through a pipe 22 inches diameter, in cubic feet, also in miner's inches?

Answer: In this table find in first column 52.8 feet, opposite which in column headed 22 Inches will be found the required quantity, viz., 21.06 cubic feet, which multiplied by 50 gives 1053 miner's inches.

Question: The diameter of the pipe being 24 inches, what fall will be required for the pipe to carry 1000 miner's inches?

Answer: In this table, in column headed 24 Inches, find that number which multiplied by 50 will make the 1000 miner's inches given. In this case the nearest number is 20.42, opposite which in column headed Fall per Mile will be found 31.68 feet, the fall required.

Question: In carrying 1050 inches of water to a hydraulic mine in a pipe 27 inches diameter, having a fall of 95.04 feet to the mile, what will be the effective head at the mine?

Answer: In this table, in column headed 27 Inches, find that number which multiplied by 50 will make 1050 approximate miner's inches. In this case we have 21.13 cubic feet, opposite which in column headed Fall per Mile we find 18.48 feet, which is the head per mile lost in carrying the water. Subtracting this from the given fall or head gives the effective head. Thus 95.04 − 18.48 = 76.56 feet effective head.

Question: There being 7.5 gallons in a cubic foot, and 86,400 seconds in a day (twenty-four hours), the fall 7.39 feet per mile, how many gallons will a pipe 40 inches diameter carry per day?

Answer: In this table, in column headed 40 Inches and opposite 7.39 feet headed Fall per Mile, will be found 37.57 cubic feet flow per second. Then 37.57 × 7.5 × 86,400 = 24,345,360 gallons.

GENERAL RULE. — The velocity per second is equal to 50 times the square root of the product of the head and diameter in feet, divided by the sum of the length and 50 times the diameter of the pipe in feet.

SHORT PIPES. — This rule applies to both long and short pipes, and is approximately accurate if the diameter does not exceed two feet.

TABLE SHOWING FLOW OF WATER PER SECOND THROUGH CLEAN IRON PIPES.

Fall Per Mile. Feet.	Fall Per Rod.		Diameters.					
	Ft.	In.	½ in. Cu Ft.	¾ in. Cu. Ft.	1 in. Cu. Ft.	1¼ in. Cu. Ft.	1¾ in. Cu. Ft.	2 in. Cu. Ft.
21.12	0	0.792	......	......	......	.. ...	.. ....	.02584
26.40	0	9.990	......	......	......	......	.02014	.02924
31.68	0	1.188	......	......	......	.01460	.02270	.03274
36 96	0	1.396	......	......	......	.01583	.02426	03492
42.24	0	1.584	......	......	.00567	.01707	.02638	.03776
47.52	0	1.782	......	......	.00617	.01816	.02838	.04081
52.80	0	1.980	......	.00316	.00677	.01963	.02988	.04321
63.36	0	2.376	.00122	.00350	.00781	.02123	.03260	.04843
73.92	0	2.772	.00124	.00377	.00841	.02282	.03556	.05150
84.48	0	3.168	.00135	.00411	.00886	.02466	.03706	.05456
95.04	0	3.564	.00143	.00445	.00961	.02577	.03923	.05740
105.60	0	3.960	.00150	.00466	.00990	.02793	.04224	.06111
158.40	0	5.940	.00197	.00589	.01245	.03458	.05175	.07399
211.20	0	7.920	.00241	.00705	.01492	.04132	.06167	.08734
264.00	0	9.900	.00279	.00798	.01666	.04577	.07145	.1095
316.80	0	11.880	.00315	.00874	.01857	.05043	.07830	.1200
369.60	1	1.86	.00340	.00951	.01988	.05424	.08381	.1288
422.40	1	3.84	.00366	.01012	.02141	.05804	.08949	.1375
475.20	1	5.82	.00389	.01086	.02283	.06191	.09400	.1442
528 00	1	7.80	.00410	.01144	.02424	.06724	.10030	.1523
633.00	1	11.76	.00453	.01282	.02676	.07400	.1110	.1634
739.20	2	3.72	.00473	.01380	02890	.08020	.1200	.1748
844.00	2	7.68	.00524	.01480	.03081	.08622	.1285	.1855
950.40	2	11.64	.00559	.01567	.03276	.09225	.1372	.1955
1056.00	3	3.60	.00589	.01656	.03458	.09692	.1450	.2047
1320.00	4	1.50	.00660	.01871	.03897	.1079	.1617	.2276
1584.00	4	11.40	.00732	.02064	.04316	.1187	.1773	.2483
2112.00	6	7.20	.00855	.02390	.04987	.1380	.2050	.2833
2640.00	8	3.00	.00966	.02705	.05648	.1550	......	.....
3168.00	9	10.80	.01065	.03003	.06320	......	......	.....
3696.00	11	6.60	.01156	.03301	.06943	......	......	.....
4224.00	13	2.40	.01248	.03572	......	......	......	.....
4752.00	14	10.20	.01338	.03786	......	......	......	.....
5280.00	16	5.00	.01419	......	......	......	......	.....

TABLE SHOWING FLOW OF WATER PER SECOND THROUGH
CLEAN IRON PIPES.

Fall Per Mile. Feet.	Fall Per Rod. Ft.	In.	Diameters.						
			3 in. Cu. Ft	4 in. Cu. Ft.	6 in. Cu. Ft.	8 in. Cu. Ft.	10 in. Cu. Ft.	11 in. Cu. Ft.	12 in Cu. Ft.
5.280	0	0.198	.....	.....	.....	.....	.....	.....	1.265
6.336	0	0.238	.....	.....	.....	.....	.878	1.120	1.402
7.392	0	0.277	.....	.....	.....	.....	.960	1.221	1.489
8.448	0	0.317	.....	.....	.....	.573	1.047	1.320	1.634
9.504	0	0.356	.....	.....	.....	.611	1.110	1.394	1.728
10.560	0	0.396	.....	.....	.298	.639	1.194	1.490	1.826
11.616	0	0.436	. ....	.....	.314	.659	1.265	1.580	1.940
12.672	0	0.475	.....	.....	.330	.703	1.325	1.653	2.026
13.728	0	0.515	.....	.1235	.346	.737	1.377	1.722	2.117
14.784	0	0.554	.....	.1298	.359	.768	1.423	1.788	2.207
15.840	0	0.594	.0630	.1335	.377	.808	1.470	1.854	2.297
18.480	0	0.684	.0692	.1465	.395	.876	1.587	1.996	2.466
21.120	0	0.792	.0749	.1562	.444	.931	1.683	2.136	2.662
26.400	0	0.990	.0839	.1771	.496	1.045	1.865	2.397	3.020
31.6?0	0	1.188	.0915	.1923	.548	1.575	2.059	2.636	3.310
36.960	0	1.386	.0992	.2146	.589	1.262	2.222	2.858	3.601
42.240	0	1.584	.1060	.2339	.631	1.314	2.383	3.062	3.856
47.520	0	1.782	.1119	.2460	.672	1.424	2.514	3.232	4.072
52.800	0	1.980	.1190	.2582	.721	1.496	2.662	3.419	4.305
63.360	0	2.376	.1313	.2893	.784	1.644	2.932	3.760	4.728
73.920	0	2.772	.1413	.3036	.858	1.782	3.210	4.016	5.094
84.480	0	3.168	.1507	.3237	.922	1.916	3.450	4.390	5.482
95.040	0	3.564	.1590	.3412	.975	2.033	3 679	4.679	5 839
105.600	0	3.960	.1717	.3607	1.022	2.155	3.856	5.251	6.160
158.400	0	5.940	.2081	.4503	1.263	2.667	4.762	6.086	7.630
211.200	0	7.920	.2469	.5331	1.484	3.145	5.563	7.022	8.860
264.000	0	9.900	.2785	.5954	1.665	3 513	6.704	8.244	9.967
316.800	0	11.880	.3049	.6390	1.929	3.847	.....	.. ..	.....
369.000	1	1.860	.3331	.6967	1.976	4.196	.....	.....	.....
422.400	1	3.840	.3559	.7506	2.144	.....	.. ...	.....	.....
475.200	1	5.820	.3816	.7960	2.274	.....	.. ...	.....	.....
528.000	1	7.800	.4043	.9464	2.399	.....	.....	.....	.....
633.600	1	11.760	.4440	.9270	.....	.....	.....	.....	.....
739.200	2	3.720	.4977	1.0060	.....	.....	.....	.....	.....
844.800	2	7.680	.5131	1.0810	.....	.....	.....	.....	.....
950.400	2	11.640	.5436	.....	.....	.....	.....	.....	.....
1056.000	3	3.600	.5832	.....	.....	.....	.....	.....	.....
1320.000	4	1.500	.6523	.....	.....	.....	.....	.....	.....
1584.000	4	11.400	.....	.....	.....	.....	.....	.....	.....

TABLE SHOWING FLOW OF WATER PER SECOND THROUGH
CLEAN IRON PIPES—*(continued.)*

Fall per Mile. Feet.	Fall per Rod. Ft. In.	Diameters.							
		14 In. Cu. Ft.	15 In. Cu. Ft.	16 In. Cu. Ft.	18 In. Cu. Ft.	20 In. Cu. Ft.	22 In. Cu. Ft.	24 In. Cu. ft.	27 In. Cu.Ft.
2.11	0 0.08	.....	....	......	......	....	....	......	....
2.64	0 0.10	.....	......	......	......	......	......	......	8.27
3.17	0 0.12	.....	......	....	... ...	3.61	4.61	6.10	8.37
3.70	0 0.14	.....	......	2.25	3.10	4.07	5.25	6.64	9.09
4.22	0 0.16	1.71	2.05	2.43	3.27	4.35	5.62	7.13	9.48
4.75	0 0 18	1.83	2.19	2.59	3.49	4 68	6.01	7.56	10.26
5.28	0 0.20	1.91	2.30	2.72	3.66	4.92	6.32	7.95	10.74
5.81	0 0.22	2.02	2.43	2.88	3.88	5.15	6.62	8.34	11.45
6.34	0 0.24	2.11	2.54	3.02	4.06	5.40	6.94	8.75	11.93
6.86	0 0.26	2.18	2.65	3.18	4.23	5.62	7.24	9.14	12.54
7.39	0 0.28	2.27	2.75	3.28	4.40	5.82	7.51	9.47	12.96
7.92	0 0.30	2.35	2.84	3.39	4.61	6.05	7.78	9.80	13.49
8.45	0 0.32	2.44	2.94	3.49	4.75	6.27	8.03	10.13	13.98
8.98	0 0.34	2.54	2.98	3.62	4.90	6.48	8.36	10.57	14.41
9.50	0 0.36	2.59	3.11	3.69	5.03	6.65	8.55	10.77	14.81
10.03	0 0.38	2.67	3.21	3.81	5 17	6.92	8.85	11.10	15.21
10.56	0 0.40	2.72	3.29	3.92	5.30	7.05	9.07	11.43	15.63
11.62	0 0.44	2.88	3.47	4.12	5.63	7.42	9.55	12.05	16.44
12.67	0 0.48	3.02	3.63	4.32	5.87	7.79	10.01	12.01	17.23
13.73	0 0.51	3.15	3.79	4.51	6.18	8.14	10.48	13.23	18.01
14.78	0 0.55	3.29	3.95	4.68	6.38	8.48	10.91	13.79	18.75
15.84	0 0.59	3.42	4.11	4.87	6.64	8.77	11.29	14.25	19.50
18.48	0 0.69	3.62	4.46	5.31	7.17	9.49	12.25	15.50	21.13
21.12	0 0.79	3.99	4.78	5.67	7.65	10.16	13.12	16.62	22.62
26.40	0 0.99	4.46	5.37	6.39	8.66	11.43	14.78	18.71	25.34
31.68	0 1.19	4.91	5.91	7.02	9.54	12.59	16.20	20.42	27.74
36.96	0 1.39	5.37	6.45	7.66	10.33	13.66	17.53	22.05	29.96
42.24	0 1.59	5.77	6.90	8.16	11.09	14.66	18.78	23.61	31.99
47.52	0 1.78	6.11	7.31	8.64	11.71	15.54	19.93	25.07	33.97
52.80	0 1.98	6.44	7.70	9.10	12.37	16.47	21.06	26.42	35.89
63.36	0 2.38	7.00	8.39	9.95	13.65	17.99	23.07	29.03	39.76
73.92	0 2.77	7.60	9.15	10.87	14.75	19.49	24.68	31.49	43.22
84.48	0 3.17	8.17	9.81	11.63	15.84	21.03	26.97	33.90	46.57
95.04	0 3.56	8.93	10.47	12.43	16.90	22.45	29.70	36.18	48.06
105.60	0 3.96	9.26	11.09	13.14	17.85	23.56	31.15	38.45	.....
158.40	0 5.94	11.39	13.66	16.17	21.86	28.86	.....	......	.....
211.20	0 7.92	13.22	15.84	18.77	....	....	......	....	....

TABLE SHOWING FLOW OF WATER PER SECOND THROUGH CLEAN IRON PIPES—(*continued.*)

Fall Per Mile.	Fall Per Rod.		Diameters.				
Feet.	Ft.	In.	54 In. Cu. Ft.	60 In. Cu. Ft.	72 In. Cu. Ft.	84 In. Cu. Ft.	96 In. Cu. Ft.
.53	0	0.02	21.96	29.77	46.99	75.43	107.77
1.06	0	0.04	31.70	38.19	57.65	104.61	152.45
1.58	0	0.06	38.53	52.09	82.53	126.18	188.45
2.11	0	0.08	45.12	59.04	95.99	145.43	218.75
2.64	0	0.10	50.23	67.56	109 42	162.75	245.30
3.17	0	0.12	55.51	74.32	121.58	177.03	267.41
3.70	0	0.14	60.21	80.51	132.04	192 04	290.53
4.22	0	0.16	63.61	86.30	139.96	207.81	310.89
4.75	0	0.18	67.20	91.99	148.72	222.44	324.20
5.28	0	0.20	72.37	96 98	157.77	235.13	350.45
5.81	0	0.22	75.71	102.39	165.97	253.34	366.19
6.34	0	0.24	79.13	107.31	173.04	264.77	382.02
6.86	0	0.26	82.54	115.53	179.26	275.16	397.85
7.39	0	0.28	85.90	116.53	187.46	287.67	414.70
7.92	0	0.30	89.52	119.68	193.93	296.37	427.76
8.45	0	0.32	92.43	123.70	200.18	307.87	443.09
8.98	0	0.34	95.35	127.63	206.40	316.15	457.42
9.50	0	0.36	97.65	131.26	212.05	326.73	470.49
10.03	c	0.38	100.19	134.79	217.71	335.79	481.53
10.56	0	0.40	103.82	138.84	225.21	348.25	496.37
11.62	0	0.44	108.78	145.98	235.52	364.92	522.76
12.67	0	0.48	113.47	152.56	246.41	389.09	547.88
13.73	0	0.51	118.48	158 65	256.17	394.43	510.01
14.78	0	0.55	123.10	164.54	267.19	408.36	592.13
15.84	0	0.59	128.19	170.43	277.88	423.36	612.00
18.48	0	0.69	138.92	183.98	299.72	482.99	....
21.12	0	0.79	147.91	197.52	320.74	....	....
26.40	0	0.99	165.80	221.95	358.52	....	....
31.68	0	1.19	182.42	244.26	.. ....	....	....
36.96	0	1.39	190.01	.. ....	....	....	....

TABLE SHOWING FLOW OF WATER PER SECOND THROUGH CLEAN IRON PIPES—(*continued.*)

Fall per Mile.	Fall per Rod.		Diameters.					
Feet.	Ft.	In.	30 In. Cu. Ft.	33 In. Cu. Ft.	36 In. Cu. Ft.	40 In. Cu. Ft.	44 In. Cu. Ft.	48 In. Cu. Ft.
1.06	0	0.04	....	....	10.29	13 88	18.15	22.98
1.58	0	0.06	7.78	10.21	12.70	17.00	22.22	27.89
2.11	0	0.08	8.99	11.65	14.56	19.68	25.55	32.93
2.64	0	0.10	10.24	12.92	16.35	22.08	28.87	37.00
3.17	0	0.12	10.97	13.99	18.02	24.43	31.46	40.21
3.70	0	0.14	11.90	15.14	19.76	26.27	34.47	43.67
4.22	0	0.16	12.84	16.36	20.85	28.14	37.05	46.81
4.75	0	0.18	13.48	17.58	22.30	29.80	39.01	49.06
5.28	0	0.20	14.21	18.74	23.47	31.46	41.06	52.15
5.81	0	0.22	15.05	19.54	24.91	33.25	42.09	54.95
6.34	0	0.24	15.81	20.28	26.12	34.68	44.97	57.36
6.86	0	0.26	16.47	21.29	27.20	36.21	46.77	60.07
7.39	0	0.28	17.18	22.20	28.24	37.57	48.83	62.02
7.92	0	0.30	17.94	23.01	29.19	39.18	50.62	64.47
8.45	0	0.32	18.58	23.76	30.29	40.54	52.46	66.53
8.98	0	0.34	19.21	24.47	31.42	41.88	54.04	68.50
9.50	0	0.36	19.66	25.22	32.48	43.07	55.48	70.62
10.03	0	0.38	20.32	26.14	33.40	44.28	57.01	72.75
10.56	0	0.40	20.79	26.94	34.49	45.20	58.85	74.44
11.62	0	0.44	21.80	28.27	36.15	48.12	61.71	78.29
12.67	0	0.48	22.83	29.02	37.74	50.48	64.35	81.68
13.73	0	0.51	23.93	31.06	39.40	52.67	66.87	85.20
14.78	0	0.55	24.86	32.28	40.86	55.04	69.57	88.46
15.84	0	0.59	25.87	33.62	42.28	56.33	72.32	91.73
18.48	0	0.69	27.96	36.17	45.95	61.09	77.95	100.40
21.12	0	0.79	29.84	38.57	48 83	65.41	83.60	105.89
26.40	0	0.99	33.55	43.12	54.89	73.09	93.37	119.34
31.68	0	1.19	36.79	47.40	59.95	80.32	103.28	130.88
36.96	0	1.39	39.66	51.35	65.17	86.70	111.74	148.09
42.24	0	1.59	42.39	54.91	69.80	92.58	119.93	153.94
47.52	0	1.78	45.23	58.20	74.33	98.00	128.26	....
52.80	0	1.98	47.71	61.62	78.46	103.99	....	....
63.36	0	2.38	52.91	68.00	82.84	... ...	....	....
73.92	0	2.77	57.65	73.95	....	....	....	....

RELATION OF CLEAN, SLIGHTLY ROUGH, AND VERY ROUGH PIPES WITH RESPECT TO THEIR CARRYING CAPACITY.

CLEAN PIPES. — The tables, as appear by the headings, have been computed for clean pipes, in other words, smooth and straight.

SLIGHTLY ROUGH PIPES. — When the pipe is slightly rough, multiply the tabulated number for clean pipes by the decimal .886 to determine its carrying capacity.

VERY ROUGH PIPES. — If the pipe is *very rough*, multiply the tabulated number for clean pipes by the decimal .773 to determine its carrying capacity.

RELATION OF THE INLET FORMS OF PIPES WITH RESPECT TO THE COEFFICIENTS OF ENTRANCE.

COEFFICIENTS. — Of the three following forms, viz., *Bell-mouthed*, *Square-edged*, and *Square-edged* projecting into the reservoir, their coefficients will be in order .900, .836, and .734.

ANGULAR BENDS AND TABLE.

ADDITIONAL HEAD REQUIRED TO OVERCOME ONE ANGULAR BEND.

Question : The velocity being 40 feet per second, what additional head is required to overcome the resistance of an angular bend whose angle of deflection is 90 degrees?

Answer: In this table find, in column headed Velocity per Second, 40, opposite which, in column headed 90° *Head*, will be found 24.45 feet, the additional head required.

TABLE SHOWING ADDITIONAL HEAD REQUIRED TO OVERCOME THE RESISTANCE OF ONE ANGULAR BEND.

Velocity per Second. Feet.	Angles of Deflection.					
	15°Head. Feet.	30° Head. Feet.	40° Head. Feet.	60° Head. Feet.	90° Head. Feet.	120°Head. Feet.
1	.0002	.0005	.002	.006	.015	.029
2	.0010	.0019	.009	.023	.061	.116
3	.0022	.0042	.019	.051	.138	.260
4	.004	.008	.035	.090	.245	.462
5	.006	.012	.054	.141	.382	.723
6	.009	.017	.078	.204	.550	1.04
7	.012	.023	.106	.277	.749	1.42
8	.016	.030	.138	.362	.978	1.85
10	.025	.047	.216	.565	1.53	2.89
15	.056	.105	.486	1.27	3.44	6.50
20	.099	.186	.863	2.26	4.85	11.56
25	.155	.291	1.35	4.45	9.55	18.06
30	.224	.419	1.94	5.09	13.75	26.01
40	.398	.745	3.45	9.04	24.45	46.23
50	.621	1.17	5.40	14.13	38.20	73.93
75	1.40	2.62	12.14	31.79	85.95	162.5
100	2.48	4.66	21.58	56.52	152.8	289.0
150	5.59	10.48	48.57	127.2	343.7	650.2
200	9.94	18.63	86.32	226.1	611.1	1156.
300	22.36	41.92	194.20	508.7	1092.	2601.

ADDITIONAL HEAD NECESSARY TO OVERCOME THE RESISTANCE OF ONE CIRCULAR BEND.

Question: The radius of the pipe being to the radius of the bend in the ratio of 1:5, the number of degrees in the bend being 90 degrees, and the velocity 75 feet per second, what is the additional head required to overcome the resistance of the bend?

Answer: In this table, in first column, headed Velocity per Second, find 75 feet, opposite which, in column headed 1: 5, 90°, is found 6.03 feet, the required head.

Question: The radius of the pipe being to the radius of the bend in the ratio of 2: 5, the number of degrees in the bend being 120 degrees, and the velocity per second 100 feet, what is the additional head required to overcome the resistance of one bend?

Answer: In this table, opposite 100 feet velocity, will be found in column headed 2: 5, 120°, the required number, viz., 21.34 feet.

RELATIVE CARRYING CAPACITY OF OPEN CHANNELS WHOSE SECTIONAL AREAS ARE EQUAL TO EACH OTHER BUT OF DIFFERENT FORMS.

The form in which the bottom width is made equal to one of the sides, and in which the base to the perpendicular of the side slope is as 3 : 4, has been adopted as the standard form when the ground will admit, it being the simplest of construction.

The relative carrying capacity for trapezoidal form — Base: depth of slope : : 3 : 4; bottom width: depth:: 5 : 4. Coefficient of capacity, 1000.

Trapezoidal form — Base : depth of slope : : 1 : 1; bottom width = depth, .994.

Coefficients: flume, 2 : 1, .961; semi-hexagonal, 1.008; square, .925; semicircular, 1.056.

Question: The fall being 6 feet per mile, the sectional area of a *square flume* 8 square feet, what will be its carrying capacity per second?

Answer: In table showing Flow of Water in Open Channels — Base to Perpendicular of Side Slopes being as 3 : 4, in column of Fall per Mile, find the given fall 6 feet, opposite which in column headed sectn. 8.0 square feet is found 13.65 cubic feet. This multiplied by the coefficient for a square, viz., .925, gives 13.64 × .925 = 12.63 cubic feet.

Remarks. — The tables for the flow of water in open channels have been computed upon the assumption that the canals are smooth and straight.

FLOW OF WATER THROUGH NOZZLES.

Question: The head being 125 feet, how many cubic feet per second will a nozzle 4 inches in diameter discharge? How many miners' inches?

Answer: In this table find in the first column the given head, 125 feet, opposite which, in column headed 4 Inches, will be found the required quantity, viz., 7.28 cubic feet × 50 = 364 *miner's inches.*

Question: Between the inlet and the nozzles of a hydraulic pipe 3 feet in diameter the distance is five miles and the total fall 275 feet. The pipe is to carry 2000 miners' inches of water, which is to be discharged through two "Little Giants," or nozzles equal in size. What will be the loss of head by the resistance in the main pipe? What will be the size of each nozzle?

Answer: In table showing Flow of Water Through

Clean Iron Pipes find in column headed 36 Inches that number which multiplied by 50 will make 2000, the given number of *miner's inches*. In this case 40.86 approximates sufficiently near, opposite which, in column, headed Fall per Mile, is found 14.78 feet, the loss of head per mile. Multiply this by 5, the length of the pipe, and we have $14.78 \times 5 = 73.9$ feet, the loss of resistance in the pipe 5 miles long. Subtracting this from the total head, $275 - 73.9 = 201.1$ feet remaining head. Again, in the table find 200 nearest 201.1 feet in column headed Head, opposite which, in column, headed 6 Inches, is found 20.64, which multiplied by 50 gives 1.032, or approximately 1000 miner's inches, which each nozzle is required to discharge. Hence the nozzles are to be 6 inches in diameter each.

TABLE SHOWING ADDITIONAL HEAD REQUIRED TO OVERCOME THE RESISTANCE OF ONE CIRCULAR BEND.

Ratio of Radius of Pipe to Radius of Bend.

Velocity per Second. Feet.	1:5 30° Head. Feet.	1:5 60° Head. Feet.	1:5 90° Head. Feet.	1:5 120° Head. Feet.	1:5 180° Head. Feet.	2:5 30° Head. Feet.	2:5 60° Head. Feet.	2:5 90° Head. Feet.	2:5 120° Head. Feet.	2:5 180° Head. Feet.
1	.0004	.0007	.0011	.0014	.0022	.0005	.001	.002	.002	.005
2	.0014	.0029	.0043	.0058	.0086	.0021	.004	.006	.008	.013
3	.0032	.0064	.0096	.0128	.0192	.0048	.010	.014	.020	.029
4	.0057	.0114	.0171	.0228	.0342	.0085	.017	.025	.034	.051
5	.0089	.0179	.0268	.0358	.0536	.0133	.027	.040	.054	.080
6	.0129	.0257	.0386	.0514	.0772	.0192	.038	.058	.076	.115
7	.0175	.0350	.0525	.0700	.1050	.0261	.052	.078	.104	.157
8	.0229	.0457	.0686	.0914	.1372	.0341	.068	.102	.136	.205
10	.0357	.0714	.1072	.1428	.2144	.0533	.107	.160	.214	.320
15	.0804	.1607	.2411	.3214	.4822	.1200	.240	.360	.480	.720
20	.1429	.2858	.4287	.5716	.8574	.2130	.426	.639	.852	1.28
25	.2232	.4464	.6696	.8928	1.34	.3333	.667	1.00	1.33	2.00
30	.3214	.6428	.9642	1.29	1.93	.4798	.960	1.44	1.92	2.88
40	.5714	1.14	1.71	2.28	3.42	.8530	1.71	2.56	3.42	5.12
50	.8927	1.79	2.68	3.58	5.36	1.33	2.66	3.99	5.32	7.98
75	2.01	4.02	6.03	8.04	12.06	3.00	6.00	9.00	12.00	18.00
100	3.57	7.14	10.71	14.28	21.42	5.33	10.67	15.99	21.34	31.98
150	8.04	16.07	24.11	32.14	48.22	12.00	24.00	36.00	48.00	72.00
200	14.29	28.58	42.87	57.16	85.74	21.32	42.64	63.96	127.92	127.92
300	32.14	64.28	96.42	128.56	192.84	47.98	95.96	143.24	287.88	287.88

FLOW OF WATER IN OPEN CHANNELS.

Question : The dimensions of a canal being, top width 11 feet, bottom width 5 feet, depth 4 feet, and the fall per mile 8 feet. Required the number of inches, miners' measure, that it will carry.

Answer : In this table, in column headed "Fall per Mile," find 8 feet, opposite which in column headed with given specifications (11, 5, 4) is found 104.8 cubic feet, the flow per second. This multiplied by 50, the number of miners' inches equal to one cubic foot flow per second, gives 104.8 × 50 = 5240 miners' inches required.

TABLE SHOWING FLOW OF WATER IN OPEN CHANNELS, BAS. TO PERPENDICULAR OF THE SIDE SLOPES BEING AS 3 : 4.

Fall per Mile. Ft.	Fall per Rod. In.	T 2.2 ft. B 1.0 ft. D .8 ft. Section 1.28 sq. ft. Cu. Ft.	T 3.3 ft. B 1.5 ft. D 1.2 ft. Section 2.88 sq. ft. Cu. Ft.	T 4.4 ft. B 2.0 ft. D 1.6 ft. Section 5.12 sq. ft. Cu. Ft.	T 5.5 ft. B 2.5 ft. D 2.0 ft. Section 8.0 sq. ft. Cu. Ft.	T 6 6 ft. B 3 0 ft. D 2 4 ft. Section 11.52 sq. ft. Cu. Ft.	T 7.7 ft. B 3.5 ft. D 2.8 ft. Section 15.68 sq. ft. Cu. Ft.	T 8.8 ft. B 4.0 ft. D 3.2 ft. Section 20.48 sq. ft. Cu. Ft.
1	.0375	.45	1.33	2.67	5.57	9.05	13.46	20.26
2	.0750	.63	1.88	3.87	7.88	12.80	19.04	28.64
3	.1125	.77	2.30	4.74	9.65	15.67	23.32	35.08
4	.15C0	.89	2.65	5.47	11.14	18.52	26.93	40.51
5	.1875	1.00	2.97	6.12	12.46	20.24	30.11	45.30
6	.2250	1.09	3.25	6.70	13.65	22.17	32.98	49.62
7	.2625	1.18	3.42	7.24	14.74	23.94	35.63	53.58
8	.3000	1.26	3.75	7.73	15.75	25.60	38.08	57.28
9	.3375	1.34	3.98	8.21	16.71	27.15	40.39	60.76
10	.3750	1.41	4.19	8.65	17.61	28.62	42.57	64.05
11	.4125	1.48	4.40	9.07	18.47	30.02	44.55	67.18
12	.4500	1.54	4.60	9.48	19.30	31.35	46.64	70.65
13	.4875	1.61	4.78	9.86	20.08	32.63	48.54	73.03
14	.5250	1.67	4.96	10.24	20.84	33.87	50.38	75.79
15	.5625	1.73	5.14	10.60	21.57	35.05	52.14	78.44
16	.6000	1.78	5.31	10.94	22.27	36.20	53.86	81.02
17	.6375	1.84	5.47	11.28	22.96	37.31	55.51	83.51
18	.6750	1.89	5.63	11.60	23.63	38.39	57.11	85.93
19	.7125	1.94	5.78	11.92	24.28	39.44	58.58	88.29
20	.7500	1.99	5.93	12.23	24.91	40.47	60.21	90.58
21	.7875	2.04	6.08	12.54	25.53	41.47	61.70	92.82
22	.8250	2.09	6.22	12.83	26.12	42.45	63.15	95.00
23	.8625	2.14	6.36	13.12	26.71	43.40	64.57	97.15
24	.9000	2 18	6.50	13.40	27.29	44.34	65.95	99.23
25	.9375	2.23	6.63	13.68	27.98	45.24	67.32	101.28

In Tables, T signifies top width; B, bottom width; D, depth.

TABLE SHOWING FLOW OF WATER IN OPEN CHANNELS,
BASE TO PERPENDICULAR OF THE SIDE SLOPES
BEING AS 3 : 4.—(*continued.*)

Fall per Mile. Ft.	Fall per Rod. In.	T 9 9 ft. B 4.5 ft. D 3.6 ft. Section 25.92 sq ft. Cu. Ft.	T 11 ft. B 5 ft. D 4 ft. Section 32 sq. ft. Cu. Ft.	T 13.2 ft. B 6.0 ft. D 4.8 ft. Section 46.09 sq. ft. Cu. Ft.	T 16.4 ft. B 7.0 ft. D 5.6 ft. Section 62.72 sq. ft. Cu. Ft.	T 17.6 ft. B 8.0 ft. D 6.4 ft. Section 81.92 sq. ft. Cu. Ft.	T 19.8 ft. B 9.0 ft. D 7.2 ft. Section 103 68 sq. in. Cu. Ft.	T 22 ft. B 10 ft. D 8 ft. Section 128 sq. ft. Cu. Ft.
1	.0375	28.04	37.1	58.4	96.5	138.3	189.2	261.2
2	.0750	39.67	52.4	82.7	136.4	195.7	267.6	369.4
3	.1125	48.59	64.2	101.4	167.1	239.6	327.7	451.3
4	.1500	56.10	74.1	117.1	192.9	276.7	378.4	522.3
5	.1875	62.71	82.9	130.9	215.7	309.3	423.1	584.0
6	.2250	68.70	90.8	143.4	236.3	338.8	463.5	639.8
7	.2625	74.19	98.1	154.8	255.3	366.0	500.5	691.0
8	.3000	79.53	104.8	165.5	272.9	391.2	535.1	738.7
9	.3375	84.14	111.1	175.6	289.4	415.0	567.6	783.5
10	.3750	88.68	117.1	185.1	305.0	437.4	598.2	825.9
11	.4125	93.02	122.9	194.1	319.9	458.7	613.2	866.2
12	.4500	97.15	128.4	202.8	334.2	479.1	655.4	925.6
13	.4875	101.13	133 6	211.1	347.8	498.7	682.1	941.7
14	.5250	104.94	138.7	219.0	360.9	517.5	707.8	977.2
15	.5625	108.63	143.5	226.6	373.6	535.7	732.8	1011.5
16	.6000	112.18	148.2	234.1	385.9	553.3	756.7	1044.7
17	.6375	115.64	152.4	241.3	397.8	570.3	780.1	1076.9
18	.6750	118.99	157.2	248.3	409.3	586.9	802.7	1108.1
19	.7125	122.26	161.5	255.1	420.5	601.5	824.8	1138.4
20	.7500	125.43	165.7	261.7	431.4	618.5	846.1	1168.0
21	.7875	128.53	169.8	268.2	442.0	633.9	867.0	1196.8
22	.8250	131.55	173.8	274.5	452.5	648.8	887 4	1225.0
23	.8625	134.51	177.7	280.7	462.9	663.4	907.4	1252.6
24	.9000	137.40	181.5	286.7	472.6	677.7	926.0	1279.5
25	.9375	140.24	185.3	292.6	482.3	691.6	946.0	1306.0

In Tables, T signifies top width; B, bottom width; D, depth.

FLOW OF WATER IN OPEN CHANNELS—(*Continued.*)

Question : Required the number of cubic feet of water that will flow in a canal whose top width is 40 feet, bottom width 20 feet, depth 5 feet, and whose fall is 2 feet per mile.

Answer : In this table, in column "Fall per Mile," find 2 feet, opposite which in column headed with the given specifications (40, 20, 5) is found the required flow, viz.,376.1 cubic feet.

TABLE SHOWING FLOW OF WATER IN OPEN CHANNELS, BASE TO PERPENDICULAR OF THE SIDE SLOPES BEING AS 2 : 1.

Fall per Mile. Feet.	Fall per Rod. Feet.	T 6 ft. B 2 ft. D 1 ft. Section 4 sq. ft. Cu. Ft.	T 9 ft. B 3 ft. D 1.5 ft. Section 9 sq. ft. Cu. Ft.	T 12 ft. B 4 ft. D 2 ft. Section 16 sq. ft. Cu. Ft.	T 16 ft. B 6 ft. D 2.5 ft. Section 27.5 sq. ft. Cu. Ft.	T 22 ft. B 10 ft. D 3 ft. Section 48 sq. ft. Cu. Ft.	T 28 ft. B 12 ft. D 4 ft. Section 30 sq. ft. Cu. Ft.	T 40 ft. B 20 ft. D 5 ft. Section 150 sq. ft. Cu. Ft.
.5	.01875	1.27	3.85	8.63	18.11	8.79	78.2	188.1
.6667	.0250	1.46	4.44	9.96	20.91	44.79	90.3	217.2
.8333	.03125	1.63	4.96	11.14	23.38	50.08	101.0	242.8
1	.0375	1.79	5.44	12.20	25.61	54.86	110.6	266.0
1.25	.046875	2.00	6.08	13.64	28.68	61.32	123.7	297.4
1.5	.05625	2.19	6.67	14.96	31.34	67.26	135.7	326.1
1.75	.065625	2.37	7.19	16.14	33.88	72.57	146.4	351.8
2	.0750	2.53	7.69	17.26	36.22	77.58	156.5	376.1
2.25	.084375	2.68	8.16	18.30	38.42	82.29	165.9	399.0
2.5	.09375	2.83	8.60	19 29	40.50	86.72	174.9	420.6
3	.1125	3.10	9.42	21.14	44.36	95.00	191.6	460.7
3.5	.13125	3.35	10.17	22.83	47.91	102.6	207.0	497.6
4	.1500	3.58	10.87	24.41	51.22	109.7	221.3	531.9
4.5	.16875	3.79	11.54	25.88	54.33	116.3	234.7	564.2
5	.1875	4.00	12.16	27.29	57.27	122.7	247.4	594.8
6	.2250	4.38	13.31	29.89	62.74	134.4	271.0	651.5
7	.2625	4.73	14.39	32.29	67.79	145.1	292.7	703.6
8	.3000	5.06	15.38	34.52	72.43	155.2	312.9	752.2
9	.3375	5.37	16.31	36.61	76.83	164.6	331.9	797.9
10	.3750	5.66	17.19	38.59	80.99	173.5	349.9	841 1
11	.4125	5.93	18.03	40.47	84.94	181.9	366.9	882.1
12	.4500	6.20	18.74	42.27	88.72	190.1	383.2	921.5

In Tables, T signifies top width ; B, bottom width ; D, depth.

TABLE SHOWING FLOW OF WATER THROUGH NOZZLES—
QUANTITY AND HORSE-POWER.

| Head | Velocity per Sec. | 50 Min. 1 Cubic Foot | 100 Min. 2 Cubic Feet | Diameters of Nozzles. | | | | | | | |
| | | | | 1 Inch. | | 1.5 Inches. | | 2 Inches. | | 2.5 Inches. | |
Feet.	Feet.	H.P.	H.P.	Cubic Feet.	H.P.	Cubic Feet.	H.P.	Cubic Feet.	H.P.	Cubic Feet.	H.P.
1	8.025	.106	.212	.041	.0046	.093	.010	.164	.018	.255	.029
1.5	9.83	.158	.316	.050	.0085	.111	019	.200	.034	.312	.053
2	11.35	.211	.422	.058	.013	.130	.029	.232	.052	.360	.082
2.5	12.68	.264	.528	.064	.018	.145	.041	.256	.072	.402	.114
3	13.90	.317	.634	.061	.024	.159	.054	.284	.096	.440	.150
3.5	15.01	.370	.740	.016	.030	.171	.068	.304	.120	.475	.189
4	16.05	.421	.842	.081	.03	.183	.083	.324	.148	.507	.231
4.5	17.02	.474	.948	.086	.044	.194	.099	.344	.176	.540	.275
5	17.95	.528	1.06	.091	.051	.205	.113	.364	.204	.56	.315
6	19.66	.634	1.27	.100	.068	.224	.153	.400	.272	.622	.425
7	21.23	.739	1.48	.108	.086	.242	.193	.432	.344	.672	.535
7.5	21.98	.702	1.58	.111	.095	.250	.214	.444	.380	.697	.595
10	25.38	1.06	2.12	.129	.146	.290	.329	.516	.584	.805	.915
12.5	28.37	1.32	2.64	.144	.204	.324	.46	.566	.816	.897	1.28
15	31.08	1.59	3.18	.158	.269	.355	.505	.632	1.08	.985	1.68
17.5	33.57	1.85	3.70	.170	.339	.383	.782	.680	1.36	1.06	2.11
20	35.89	2.11	4.22	.182	.414	.410	.931	.728	1.66	1.14	2.58
22.5	38.07	2.38	4.76	.193	.494	.435	1.11	.772	1.98	1.21	3.08
25	40.13	2.64	5.28	.204	.578	.458	1.30	.816	2.31	1.27	3.61
27.5	42.08	2.90	5.80	.213	.660	.480	1.50	.852	2.60	1.33	4.17
30	43.95	3.02	6.04	.228	.760	.513	1.71	.912	3.04	1.42	4.75
32.5	45.75	3.34	6.68	.232	.857	.522	1.93	.928	3.43	1.45	5.35
35	47.47	3.69	7.38	.241	.958	.542	2.15	.964	3.83	1.51	5.98
40	50.75	4.22	8.44	.257	1.17	.579	2.63	1.03	4.68	1.61	7.31
45	53.83	4.75	9.50	.273	1.40	.614	3.14	1.09	5.60	1.71	8.23
50	56.75	5.28	10.56	.288	1.64	.648	3.68	1.15	6.56	1.79	10.22
60	62.16	6.34	12.68	.385	2.15	.709	4.84	1.26	8.60	1.97	13.43
70	67.14	7.39	14.78	.341	2.71	.766	6.10	1.36	10.84	2.13	16.93
80	71.78	8.46	16.90	.364	3.31	.819	7.45	1.46	13.24	2.27	20.69
90	76.13	9.53	19.06	.386	3.95	.864	8.88	1.54	15.80	2.44	24.68
100	80.25	10.56	21.12	.407	4.63	.916	10.41	1.63	18.52	2.54	28.90
125	89.72	13.21	26.42	.455	6.47	1.02	14.55	1.82	25.88	2.81	40.40
150	98.28	15.85	31.70	.499	8.50	1.12	19.12	2.00	34.00	3.11	53.12
175	106.1	18.50	37.00	.539	10.70	1.21	24.07	2.16	42.80	3 36	66.86
200	113.5	21.14	42.28	.576	13.1	1.29	29.43	2.30	52.4	3.50	81.75
250	127.1	26.62	52.84	.644	18.3	1.45	41.13	2.58	73.2	4.02	114.2
300	139.0	31.70	63.40	.705	24.0	1.59	54.07	2.82	96.0	4.40	150.2
350	150.1	37.08	74.16	.762	30.3	1.71	68.15	3.05	121.2	4.76	189.3
400	160.5	42.27	84.54	.814	37.0	1.83	83.25	3.26	148.0	5.09	231.2
450	170.2	47.64	95.28	.864	44.2	1.94	99.34	3.46	176.8	5.40	276.0
500	179.4	52.84	105.7	.910	51.7	2.05	116.5	3.64	206.8	5.60	323.2
550	188.2	58.22	116.4	.955	59.7	2.10	134.2	3.82	238.8	5.96	372.7
600	196.6	63.41	126.8	.999	68.0	2.23	152.9	3.99	272.0	6.23	475.0
700	212.3	73.98	148.0	1.06	85.7	2.46	192.8	4.36	342.8	6.79	535.5
800	226.9	84.55	169.1	1.15	104.7	2.58	235.5	4.60	418.8	7.19	654.0
900	240.7	95.14	190.3	1.22	124.9	2.75	281.0	4.88	499.6	7.63	780.5
1000	253.8	105.6	211.2	1.29	146.2	2.89	329.0	5.16	584.8	8.04	914.0

TABLE SHOWING FLOW OF WATER THROUGH NOZZLES—
QUANTITY AND HORSE-POWER—(continued.)

Head	Velocity per Sec.	150 Min. Inch. 3 Cubic Feet.	200 Min. Inch. 4 Cubic Feet.	Diameters of Nozzles.							
				3 Inches.		3.5 Inches.		4 Inches.		4.5 Inches.	
Feet.	Feet.	H.P.	H.P.	Cubic Feet.	H.P.	Cubic Feet.	H.P.	Cubic Feet.	H.P.	Cubic Feet.	H.P.
1	8.025	.308	.424	.372	.040	.50	.056	.656	.072	.81	.090
1.5	9.83	.474	.632	.444	.076	.61	.105	.800	.136	1.00	.171
2	11.35	.633	.844	.520	.116	.70	.160	.928	.208	1.17	.260
2.5	12.68	.792	1.06	.58	.164	.79	.224	1.02	.288	1.30	.370
3	13.90	.951	1.27	.636	.216	.86	.295	1.14	.384	1.43	4.85
3.5	15.01	1.110	1.48	.684	.272	.94	.370	1.22	.480	1.54	.612
4	16.05	1.26	1.68	.742	.332	1.02	.452	1.30	.592	1.64	.742
4.5	17.02	1.42	1.90	.776	.396	1.06	.540	1.38	.704	1.74	.815
5	17.95	1.58	2.12	.820	.452	1.11	.600	1.46	.816	1.84	1.02
6	19.66	1.90	2.54	.896	.612	1.22	.833	1.60	1.09	2.01	1.38
7	21.23	2.22	2.96	.968	.772	1.32	1.05	1.73	1.38	2.18	1.74
7.5	21.98	2.38	3.16	1.00	.856	1.36	1.16	1.78	1.52	2.25	1.92
10	25.38	3.18	4.24	1.16	1.32	1.57	1.79	2.16	2.34	2.61	2.97
12.5	28.37	3.96	5.28	1.30	1.84	1.76	2.50	2.30	3.46	2.92	4.14
15	31.08	4.77	6.36	1.42	2.42	1.93	3.29	2.53	4.32	3.19	5.44
17.5	33.57	5.55	7.40	1.53	3.13	2.08	4.20	2.72	5.44	3.44	7.04
20	35.89	6.33	8.44	1.64	3.72	2.23	5.07	2.91	6.64	3.69	8.37
22.5	38.07	7.14	9.52	1.74	4.44	2.36	6.05	3.09	7.92	3.91	9.99
25	40.13	7.92	10.56	1.83	5.20	2.54	7.08	3.26	9.24	4.12	11.70
27.5	42.08	8.70	11.60	1.92	6.00	2.61	8.17	3.41	10.68	4.32	13.50
30	43.95	9.06	12.08	2.05	6.84	2.79	9.31	3.65	12.16	4.61	15.39
32.5	45.75	10.02	13.36	2.09	7.72	2.84	10.50	3.71	13.72	4.70	17.37
35	47.47	11.07	14.76	2.17	8.60	2.95	11.71	3.86	15.32	4.88	19.35
40	50.75	12.66	16.88	2.32	10.52	3.15	14.33	4.12	18.72	5.22	23.67
45	53.83	14.25	19.00	2.46	12.56	3.34	17.10	4.36	22.40	5.54	28.25
50	56.75	15.84	21.12	2.59	14.72	3.52	20.03	4.60	26.24	5.83	32.12
60	62.10	19.20	25.36	2.84	19.36	3.86	26.32	5.04	34.40	6.39	43.55
70	67.14	22.17	29.56	3.06	24.40	4.17	33.17	5.42	43.36	6.84	54.90
80	71.78	25.36	33.86	3.28	20.80	4.40	40.55	5.81	52.96	7.38	67.05
90	76.13	28.59	38.12	3.46	35.52	4.73	48.37	6.16	68.20	7.78	79.92
100	80.25	31.68	42.24	3.66	41.64	4.98	56.67	6.52	74.08	8.23	93.70
125	89.72	39.63	52.84	4.08	58.20	5.57	79.20	7.28	103.5	9.18	130.9
150	98.28	47.55	63.40	4.48	76.48	6.10	104.10	8.00	136.0	10.08	172.1
175	106.1	55.50	74.00	4.84	96.28	6.60	131.07	8.04	171.2	10.89	216.6
200	113.5	63.42	84.56	5.10	117.7	7.05	160.22	9.20	219.6	11.61	261.7
250	127.1	79.26	105.7	5.80	164.5	7.88	223.92	10.32	292.8	13.05	370.2
300	139.0	95.10	126.8	6.36	216.3	8.63	294.3	11.28	384.0	14.31	486.9
350	150.1	111.7	148.3	6.84	272.6	9.33	371.2	12.20	484.8	15.39	613.2
400	160.5	126.8	169.1	7.32	323.0	9.97	453.2	13.04	592.0	16.47	749.2
450	170.2	142.9	190.6	7.76	397.4	10.58	541.0	13.84	707.2	17.46	894.2
500	179.4	158.5	211.4	8.20	406.0	11.15	627.0	14.56	827.2	18.45	104.8
550	188.2	174.7	232.8	8.40	536.8	11.69	731.0	15.28	955.2	18.90	1208
600	196.6	190.2	253.6	8.92	611.0	12.21	832.7	15.96	1080.0	20.07	1376
700	212.3	221.9	296.0	9.84	771.2	13.31	1051	17.44	1371.2	22.14	1735
800	226.9	253.6	338.2	10.32	942.0	14.10	1282	18.40	1675.2	23.22	2119
900	240.7	285.4	380.6	11.00	1124	14.9	1530	19.52	1998.4	24.75	2529
1000	253.8	316.8	422.4	11.56	1316	15.76	1791	30.64	2339.2	26.00	2961

TABLE SHOWING FLOW OF WATER THROUGH NOZZLES—
QUANTITY AND HORSE-POWER—(*continued*.)

Head	Velocity per Sec.	300 Min. Inch 6 Cubic Feet	400 Min. Inch 8 Cubic Feet	Diameters of Nozzles.							
				5 Inches.		5.5 Inches.		6 Inches.		7 Inches.	
Feet.	Feet.	H.P.	H.P.	Cubic Feet.	H.P.	Cubic Feet.	H.P.	Cubic Feet.	H.P.	Cubic Feet.	H.P.
1	8.025	.616	8.8	1.02	.116	1.23	.140	1.49	.100	1.99	.226
1.5	9.83	.948	1.26	1.25	.212	1.51	.257	1.78	.304	2.44	.420
2	11.35	1.27	1.69	1.44	.327	1.74	.395	2.08	.464	2.82	.641
2.5	12.68	1.58	2.11	1.61	.457	1.95	.553	2.32	.656	3.15	.896
3	13.90	1.90	2.54	1.76	.601	2.13	.727	2.54	.864	3.45	1.18
3.5	15.01	2.22	2.96	1.90	.757	2.31	.916	2.74	1.09	3.78	1.48
4	16.05	2.53	3.37	2.03	.925	2.46	1.12	2.97	1.33	4.09	1.81
4.5	17.02	2.84	3.79	2.16	1.10	2.51	1.33	3.10	1.58	4.23	2.16
5	17.95	3.18	4.24	2.27	1.26	2.75	1.53	3.28	1.81	4.40	2.48
6	19.66	3.81	5.08	2.49	1.70	3.02	2.05	3.58	2.45	4.88	3.33
7	21.23	4.44	5.92	2.69	2.14	3.26	2.55	3.87	3.09	5.28	4.20
7.5	21.98	4.74	6.32	2.79	2.38	3.42	2.87	4.00	3.42	5.40	4.66
10	25.38	6.36	8.48	3.22	3.66	3.89	4.42	4.64	5.28	6.30	7.16
12.5	28.37	7.92	10.56	3.59	5.11	4.3	6.18	5.20	7.36	7.05	10.02
15	31.08	9.54	12.72	3.94	6.72	4.76	8.13	5.68	8.08	7.72	13.17
17.5	33.57	11.10	14.80	4.26	8.46	5.15	10.24	6.12	12.52	8.34	16.80
20	35.80	12.66	16.88	4.55	10.34	5.50	12.51	6.56	14.88	8.92	20.28
22.5	38.07	14.28	19.04	4.83	12.34	5.84	14.93	6.96	17.76	9.46	24.20
25	40.13	15.84	21.12	5.09	14.45	6.16	17.49	7.32	20.80	10.15	28.33
27.5	42.08	17.40	23.20	5.34	16.67	6.46	20.18	7.68	24.00	10.40	32.08
30	43.95	18.12	24.16	5.70	19.00	6.90	22.99	8.20	27.36	11.18	37.25
32.5	45.75	20.04	26.72	5.80	21.42	7.02	25.92	8.36	30.88	11.37	41.99
35	47.47	22.14	29.52	6.02	23.94	7.28	28.97	8.68	33.40	11.80	46.84
40	50.75	25.32	33.76	6.44	29.25	7.78	83.39	9.28	42.08	12.61	57.33
45	53.83	29.50	38.00	6.82	34.90	8.26	42.23	9.84	50.24	13.38	68.40
50	56.75	31.68	42.24	7.19	40.87	8.70	49.46	10.36	58.88	14.10	80.11
60	62.16	38.04	50.72	7.88	53.72	9.54	65.01	11.36	77.44	15.44	105.3
70	67.14	44.34	59.12	8.51	67.72	10.30	81.95	12.24	97.60	16.09	132.7
80	71.78	50.74	67.64	9.10	82.76	11.01	100.1	13.12	119.2	17.84	162.2
90	76.13	57.18	76.24	9.65	98.72	11.58	119.5	13.84	142.1	18.92	193.5
100	80.25	63.36	84.48	10.17	115.6	12.31	139.9	14.64	166.6	19.94	226.7
125	89.72	79.26	95.68	11.38	161.6	13.76	195.0	16.32	232.8	22.30	316.8
150	98.28	95.10	126.8	12.46	212.5	15.08	257.0	17.92	305.9	24.42	416.4
175	106.1	111.0	148.0	13.46	267.5	15.29	313.7	19.36	385.1	26.39	524.3
200	113.5	126.8	169.1	14.34	327.0	17.51	395.7	20.64	470.8	28.20	640.9
250	127.1	158.5	211.4	16.09	457.0	19.47	553.0	23.20	658.0	31.54	855.7
300	139.0	190.2	253.6	17.62	601.0	21.33	726.9	25.44	865.2	34.54	1177
350	150.1	222.5	296.6	19.04	757.2	22.04	916.3	27.36	1090.4	37.32	1485
400	160.5	253.6	338.2	20.35	925.0	24.62	1179	29.28	1332	39.89	1813
450	170.2	285.8	381.1	21.59	1104	26.12	1335	31.04	1590	42.31	2164
500	179.4	317.1	422.8	22.75	1293	27.54	1565	32.80	1864	44.00	2508
550	188.2	349.2	465.6	23.86	1491	28.88	1805	33.60	2147	46.78	2923
600	196.0	380.4	507.2	24.93	1699	30.16	2056	35.08	2446	48.86	3331
700	212.3	444.0	592.0	27.18	2142	32.88	2591	39.36	3085	53.26	4203
800	226.9	507.3	676.4	28.77	2616	34.92	3166	41.28	3768	56.40	5129
900	240.7	570.9	761.2	30.52	3122	36.94	3778	44.00	4496	59.83	6120
1000	253.8	633.6	844.8	32.17	3656	38.93	4424	46.24	5264	63.06	7166

TABLE SHOWING FLOW OF WATER THROUGH NOZZLES—
QUANTITY AND HORSE-POWER.

Head Feet.	Velocity per Sec. Feet.	500 Min. Inches. 10 Cubic Feet. H.P.	1000 Min. Inches. 20 Cubic Feet. H.P.	Diameters of Nozzles.							
				8 Inches.		9 Inches.		10 Inches.		12 Inches.	
				Cubic Feet.	H.P.	Cubic Feet.	H.P.	Cubic Feet.	H.P.	Cubic Feet.	H.P.
1	8.025	1.06	2.12	2.62	.288	3.35	.360	4.07	.46	5.96	.904
1.5	9.83	1.58	3.16	3.20	.544	3.99	.684	4.99	.85	7.12	1.68
2	11.35	2.11	4.22	3.71	.832	4.68	1.04	5.76	1.3	8.32	2.56
2.5	12.68	2.64	5.28	4.08	1.15	5.22	1.48	6.44	1.83	9.28	3.58
3	13.90	3.17	6.34	4.56	1.54	5.72	1.94	7.05	2.40	10.16	4.72
3.5	15.01	3.70	7.40	4.88	1.92	6.16	2.45	7.62	3.03	10.96	5.92
4	16.05	4.21	8.42	5.20	2.37	6.58	2.99	8.14	3.70	11.88	7.24
4.5	17.02	4.74	9.48	5.52	2.81	6.98	3.26	8.64	4.42	12.40	8.64
5	17.95	5.28	10.6	5.84	3.26	7.38	4.07	9.10	5.05	13.12	9.92
6	19.66	6.34	12.7	6.40	4.36	8.06	5.51	9.97	6.80	14.32	13.32
7	21.23	7.39	14.8	6.92	5.52	8.71	6.95	10.77	8.57	15.48	16.80
7.5	21.98	7.92	15.8	7.12	6.08	9.00	7.70	11.14	9.50	16.00	18.64
10	25.38	10.6	21.2	8.64	9.36	10.41	11.88	12.87	14.63	18.56	28.64
12.5	28.37	13.2	26.4	9.20	13.84	11.70	16.56	14.39	20.44	20.80	40.08
15	31.08	15.9	31.8	10.12	17.28	12.78	21.78	15.76	26.87	22.72	52.68
17.5	33.57	18.5	37.0	10.88	21.76	13.77	28.17	17.03	33.86	24.48	67.20
20	35.89	21.1	42.2	11.64	26.56	14.76	33.48	18.20	41.37	26.24	81.12
22.5	38.07	23.8	47.6	12.36	31.68	15.66	39.96	19.31	49.37	27.84	96.80
25	40.13	26.4	52.8	13.04	36.96	16.47	46.80	20.35	57.82	29.28	113.3
27.5	42.08	29.0	58.0	13.64	42.72	17.28	54.00	21.34	66.70	30.72	130.7
30	43.95	30.2	60.4	14.60	48.64	18.45	61.56	22.81	76.01	32.80	149.0
32.5	45.75	33.4	66.8	14.84	54.88	18.81	69.48	23.20	85.70	33.44	168.9
35	47.47	36.9	73.8	15.44	61.28	19.53	77.40	24.08	95.78	34.72	187.4
40	50.75	42.2	84.4	16.48	74.88	20.88	94.68	25.74	117.0	37.12	229.3
45	53.83	47.5	95.0	17.44	89.60	22.14	113.0	27.30	139.6	39.36	273.6
50	56.75	52.8	105.6	18.40	105.0	23.31	128.5	28.78	163.5	41.44	320.4
60	62.16	63.4	126.8	20.16	137.6	25.56	174.2	31.53	214.9	45.44	421.2
70	67.14	73.9	147.8	21.68	173.4	27.54	219.6	34.06	270.9	48.96	530.8
80	71.78	84.6	169.0	23.36	211.8	29.52	268.2	36.41	331.0	52.48	648.8
90	76.13	95.3	190.6	24.64	252.8	31.14	319.7	38.61	394.9	55.36	774.0
100	80.25	106.6	211.2	26.08	296.3	32.94	374.8	40.70	462.5	58.56	906.8
125	89.72	132.1	264.2	29.12	414.0	36.72	523.8	45.51	646.5	65.28	1267
150	98.28	158.5	317.0	32.00	554.0	40.32	688.3	49.85	849.8	71.68	1666
175	106.1	185.0	370.0	34.56	684.8	43.56	866.5	53.85	1070	77.44	2097
200	113.5	211.4	422.8	36.80	878.4	46.44	1059	57.56	1308	82.56	2564
250	127.1	264.2	528.4	41.28	1171	52.20	1481	64.36	1828	92.80	3583
300	139.0	317.0	634.0	45.12	1536	57.24	1947	70.50	2403	101.76	4708
350	150.1	370.8	741.6	48.80	1949	61.56	2453	76.15	3029	109.4	5940
400	160.5	422.7	845.4	52.16	2368	65.88	2997	81.41	3700	117.1	7252
450	170.2	476.4	952.8	55.36	2829	69.84	3577	86.35	4415	124.2	8656
500	179.4	528.4	1057	58.24	3409	73.80	4194	91.02	5172	131.2	10032
550	188.2	582.2	1164	61.12	3821	75.60	4831	95.46	5966	134.4	11692
600	196.6	634.1	1268	63.84	4352	80.28	5504	99.71	6798	142.7	13324
700	212.3	739.8	1480	69.76	5485	88.56	6191	108.7	8567	157.4	16812
800	226.9	845.5	1691	73.60	6701	92.88	8478	115.1	10468	165.1	20516
900	240.7	951.4	1903	78.08	7994	99.00	10116	122.5	12489	176.0	24480
1000	253.8	1056	2112	82.56	9357	104.0	11844	128.7	14624	185.0	28664

Explanation of Pipe Tables.

The tables for sheet-iron pipe are arranged as follows:

Column No. 1 gives the diameter of the pipe in inches.

Column No. 2 is the area in square inches corresponding to the diameter.

Column No. 3 is the thickness of the iron or steel in decimal parts of an inch.

Column No. 4 is the thickness of the iron or steel by the Birmingham wire gauge.

Column No. 5 is the working pressure the pipe will be subjected to in pounds per square inch, allowing 10,000 pounds tensile strain per sectional inch of iron, deducting 25 per cent for riveted joints.

For steel pipes use 14,000 pounds tensile strain per sectional inch; deduct 25 per cent for riveted joints. Hence, *working pressure for steel pipe* may be taken 40 per cent higher than given in table.

Column No. 6 is the number of cubic feet of water that will flow through the pipe in one minute, when the velocity of the water is 10 feet per second.

Column No. 7 is an approximation of the weight of a lineal foot of pipe, including rivets.

The cost of pipe varies with the iron market, and the quantity ordered of one diameter and thickness — small lots costing sometimes 50 per cent more than large orders.

The charge is extra for dipping pipes in asphaltum — coating them inside and outside — and the cost of dipping small pipes is about one cent for each inch in diameter and one foot in length.

Coating with asphaltum adds about one third of a pound per square foot to the pipe.

Diameter of Pipe.	Area of Pipe.	Thickness of Iron in Inches.	Thickness of Iron by Wire Gauge.	Pressure the Pipe will stand.	Cub. Ft. discharged per minute at a velocity of 10 ft. per second.	Weight of Pipe per Lineal Foot.
1	2	3	4	5	6	7
3	7	0.035	20	176	29	$1\frac{1}{2}$
3	7	0.049	18	245	29	$2\frac{1}{4}$
4	12.5	0.049	18	183	52	3
4	12.5	0.065	16	243	52	4
5	19.6	0.049	18	147	81	$3\frac{1}{2}$
5	19.6	0.065	16	195	81	5
5	19.6	0.083	14	249	81	$6\frac{1}{4}$
5	19.6	0.109	12	327	81	$8\frac{1}{2}$
6	28	.049	18	122	116	$4\frac{1}{4}$
6	28	.065	16	162	116	$5\frac{3}{4}$
6	28	.083	14	207	116	$7\frac{1}{3}$
6	28	.109	12	272	116	10
7	38	.049	18	105	158	$5\frac{1}{4}$
7	38	.065	16	141	158	$6\frac{3}{4}$
7	38	.093	14	178	158	$8\frac{1}{2}$
7	38	.109	12	234	158	$11\frac{1}{2}$
8	50	.065	16	119	208	$7\frac{1}{2}$
8	50	.083	14	555	208	$9\frac{1}{3}$
8	50	.109	12	204	208	13
8	50	.120	11	225	208	14
8	50	.134	10	252	208	$15\frac{3}{4}$
9	63	.065	16	108	262	$8\frac{1}{4}$
9	63	.083	14	138	262	$10\frac{3}{4}$
9	63	.109	12	182	...	$14\frac{1}{4}$
9	63	.120	11	200	262	16
9	63	.134	10	228	262	$17\frac{1}{2}$
10	78	.065	16	98	313	$9\frac{1}{4}$
10	78	.083	14	125	313	$11\frac{3}{4}$
10	78	.109	12	164	313	$15\frac{1}{2}$
10	78	.120	11	...	313	$17\frac{1}{2}$
10	78	.134	10	201	313	$19\frac{1}{4}$
11	95	.065	16	89	378	$9\frac{3}{4}$
11	95	.083	14	113	378	13
11	95	.109	12	149	378	$17\frac{1}{4}$
11	95	.120	11	162	378	$18\frac{3}{4}$
11	95	.134	10	183	378	21
12	113	.065	16	81	470	$11\frac{1}{4}$
12	113	.083	14	104	470	14
12	113	.109	12	136	470	$18\frac{1}{2}$

1 Diameter of Pipe.	2 Area of Pipe.	3 Thickness of Iron in Inches.	4 Thickness of Iron by Wire Gauge.	5 Pressure the Pipe will stand.	6 Cub. Ft. discharged per minute at a velocity of 10 ft. per second.	7 Weight of Pipe per Lineal Foot.
12	113	.120	11	150	470	19¾
12	113	.134	10	168	470	22¾
12	113	.165	8	206	470	27¾
12	113	.180	7	225	470	32
12	113	.203	6	254	470	36
12	113	.238	4	297	470	41½
13	132	.065	16	75	550	12
13	132	.083	14	93	550	15
13	132	.109	12	114	550	20
13	132	.120	11	138	550	22
13	132	.134	10	155	550	24½
13	132	.165	8	190	550	30
13	132	.180	7	207	550	34½
13	132	.203	6	234	550	38½
13	132	.238	4	275	550	42½
14	153	.065	16	69	637	13
14	153	.083	14	89	637	16
14	153	.109	12	117	637	21½
14	153	.120	11	129	637	23½
14	153	.134	10	144	637	26
14	153	.165	8	177	637	32
14	153	.180	7	193	637	37
14	153	.203	6	217	637	41¼
14	153	.238	4	255	637	48
15	176	.065	16	65	733	13¾
15	176	.083	14	83	733	17
15	176	.109	12	109	733	23
15	176	.120	11	120	733	24½
15	176	.134	10	134	733	28
15	176	.165	8	165	733	34¼
15	176	.180	7	180	733	39
15	176	.203	6	203	733	43¾
15	176	.238	4	238	733	51
16	201	.065	16	61	837	14½
16	201	.083	14	78	837	17¼
16	201	.109	12	102	837	24¼
16	201	.120	11	113	837	26½
16	201	.134	10	126	837	29½
16	201	.165	8	155	837	36

Diameter of Pipe. 1	Area of Pipe. 2	Thickness of Iron in Inches. 3	Thickness of Iron by Wire Gauge. 4	Pressure the Pipe will stand. 5	Cub. Ft. discharged per minute at a velocity of 10 ft. per second. 6	Weight of Pipe per Lineal Foot. 7
16	201	.180	7	169	837	41½
16	201	.203	6	190	837	48¼
16	201	.238	4	223	837	54½
18	254	.065	16	54	1058	16½
18	254	.083	14	69	1058	20½
18	254	.109	12	91	1058	27¼
18	254	.120	11	100	1058	30
18	254	.134	10	111	1058	34
18	254	.165	8	138	1058	41
18	254	.180	7	150	1058	46
18	254	.203	6	169	1058	51½
18	254	.238	4	198	1058	60
20	314	.065	16	49	1308	18
20	314	.083	14	63	1308	22½
20	314	.109	12	82	1308	30
20	314	.120	11	90	1308	32½
20	314	.134	10	101	1308	36½
20	314	.165	8	124	1308	44½
20	314	.180	7	135	1308	50¾
20	314	.203	6	153	1308	56¾
20	314	.238	4	179	1308	66
22	380	.065	16	45	1583	20
22	380	.083	14	57	1583	24¾
22	380	.109	12	75	1583	32¾
22	380	.120	11	82	1583	35¾
22	380	.134	10	91	1583	40
22	380	.165	8	112	1583	48¾
22	380	.180	7	123	1583	51
22	380	.203	6	138	1583	62
22	380	.238	4	162	1583	72
24	452	.083	14	52	1883	27¼
24	452	.109	12	68	1883	35½
24	452	.120	11	75	1883	39
24	452	.134	10	84	1883	43½
24	452	.165	8	103	1883	53
24	452	.180	7	112	1883	60
24	452	.203	6	127	1883	67¼
24	452	.238	4	149	1883	78¼
26	530	.083	14	48	2208	29¼

Diameter of Pipe.	Area of Pipe.	Thickness of Iron in Inches.	Thickness of Iron by Wire Gauge.	Pressure the Pipe will stand.	Cub. Ft. discharged per minute at a velocity of 10 ft. per second.	Weight of Pipe per Lineal Foot.
1	2	3	4	5	6	7
26	530	.109	12	63	2208	38$\frac{1}{2}$
26	530	.120	11	69	2208	42
26	530	.134	10	78	2208	47
26	530	.165	8	95	2208	57$\frac{1}{4}$
26	530	.180	7	104	2208	64$\frac{3}{4}$
26	530	.203	6	117	2208	72$\frac{1}{2}$
26	530	.238	4	138	2208	84
28	615	.083	14	45	2562	31$\frac{1}{4}$
28	615	.109	12	58	2562	41$\frac{1}{4}$
28	615	.120	11	64	2562	45
28	615	.134	10	72	2562	50$\frac{1}{4}$
28	615	.165	8	88	2562	61$\frac{1}{4}$
28	615	.180	7	96	2562	69$\frac{1}{2}$
28	615	.203	6	108	2562	77$\frac{1}{2}$
28	615	.238	4	127	2562	90$\frac{1}{4}$
30	706	.109	12	54	2941	44
30	706	.120	11	60	2941	48
30	706	.134	10	67	2941	54
30	706	.165	8	82	2941	65
30	706	.180	7	90	2941	74
30	706	.203	6	102	2941	83
30	706	.238	4	119	2941	96
30	706	.250	$\frac{1}{4}$	125	2941	101
33	855	.120	11	54	3561	53
33	855	.134	10	61	3561	59
33	855	.165	8	75	3561	72
33	855	.180	7	82	3561	81
33	855	.203	6	93	3561	90
33	855	.238	4	108	3561	105
33	855	.250	$\frac{1}{4}$	113	3561	110
33	855	.259	3	118	3561	115
33	855	.3125	$\frac{5}{16}$	142	3561	141
36	1017	.120	11	50	4236	58
36	1017	.134	10	56	4236	67
36	1017	.165	8	69	4236	78
36	1017	.180	7	75	4236	88
36	1017	.203	6	84	4236	98
36	1017	.238	4	92	4236	114
36	1017	.250	$\frac{1}{4}$	104	4236	120

Diameter of Pipe.	Area of Pipe.	Thickness of Iron in Inches.	Thickness of Iron by Wire Gauge.	Pressure the Pipe will stand.	Cub. Ft. discharged per minute at a velocity of 10 ft. per second.	Weight of Pipe per Lineal Foot.
1	2	3	4	5	6	7
36	1017	259	3	108	4236	125
36	1017	.3125	$\frac{5}{16}$	130	4236	153
36	1017	.375	$\frac{3}{8}$	156	4236	186
40	1256	.134	10	51	5232	71
40	1256	.165	8	62	5232	86
40	1256	.180	7	68	5232	97
40	1256	.203	6	76	5232	108
40	1256	.238	4	90	5232	126
40	1256	.250	$\frac{1}{4}$	94	5232	132
40	1256	.259	3	97	5232	138
40	1256	.3125	$\frac{5}{16}$	117	5232	169
40	1256	.375	$\frac{3}{8}$	141	5232	205
42	1385	.134	10	48	5769	$74\frac{1}{2}$
42	1385	.165	8	59	5769	91
42	1385	.180	7	64	5769	102
42	1385	.203	6	72	5769	114
42	1385	.238	4	85	5769	133
42	1385	.250	$\frac{1}{4}$	89	5769	137
42	1385	.259	3	92	5769	145
42	1385	.3125	$\frac{5}{16}$	111	5769	177
42	1385	.375	$\frac{3}{8}$	134	5769	216
44	1520	.134	10	45	6332	78
44	1520	.165	8	56	6332	95
44	1520	.180	7	61	6332	106
44	1520	.203	6	69	6332	119
44	1520	.238	4	81	6332	139
44	1520	.250	$\frac{1}{4}$	85	6332	145
44	1520	.259	3	88	6332	151
44	1520	.3125	$\frac{5}{16}$	106	6332	185
44	1520	.375	$\frac{3}{8}$	128	6332	225
48	1809	.134	10	42	7536	85
48	1809	.165	8	51	7536	103
48	1809	.180	7	56	7536	116
48	1809	.203	6	63	7536	130
48	1809	.238	4	75	7536	151
48	1809	.250	$\frac{1}{4}$	78	7536	158
48	1809	.259	3	81	7536	164
48	1809	.3125	$\frac{5}{16}$	98	7536	210
48	1809	.375	$\frac{3}{8}$	117	7536	245

FLOW OF WATER THROUGH RECTANGULAR ORIFICES
IN THIN VERTICAL PARTITIONS.

Question: The head being 10 feet, and the gate-opening being 6 inches high and 1 foot wide, what will be the discharge in miner's inches?

Answer: In this table, opposite 10 *feet* in first column, find in column headed 6 Inches High, 1 Foot Wide, 7.62 cubic feet. Multiply this number by 50, the number of miner's inches in 1 cubic foot, and there results 762 × 50 = 381.00 miner's inches.

Question: The head being 25 feet and the opening $1\frac{5}{10}$ inches high, 1 foot wide, how many pounds will be discharged per second?

Answer: In this table, opposite 25 feet in first column, find in column headed 1.5 Inches High, 1 Foot Wide, 3.05 cubic feet. Multiply this number by 62.5, the number of pounds in a cubic foot. 3.05 × 62.5 = 190.625 pounds.

Question: The head being 7 feet and the opening 1 inch high, 1 foot wide, what will be the discharge in cubic feet?

Answer: In this table, opposite 7 feet in first column, find in column headed 3 Inches High, 1 Foot Wide, 3.24 cubic feet. The given height 1 inch is one-third of 3 inches, the height of the opening; hence, without sensible error, we may take one-third the flow due 3 inches for that opening. 3.24 ÷ 3 = 1.08.

TABLE SHOWING FLOW OF WATER THROUGH RECTANGULAR ORIFICES IN THIN VERTICAL PARTITIONS.

Head upon Center of Orifice. Feet.	Velocity per Second. Feet.	1 ft. High. 1 ft. Wide.		9 in. High. 1 ft. Wide.		6 in. High. 1 ft. Wide.		3 in. High. 1 ft. Wide.		1.5 in. High. 1 ft. Wide.	
		Cubic Feet.	H.P.	Cubic Feet.	H.P.	Cubic Feet.	H.P.	Cubic Feet.	H.P.	Cubic Feet.	H.P.
0.2	3.58	....	....	....	....	....	....	....	....	.028	.0064
.3	4.40	....	....	....	....	....	....	0.69	.022	.34	.010
.4	5.07	....	....	....	....	1.56	.071	.80	.036	.40	.018
.5	5.67	....	....	2.57	.146	1.74	.099	.80	.051	.45	.026
.6	6.22	3.72	.253	2.83	.103	1.91	.130	.98	.066	.49	.033
.7	6.72	4.02	.317	3.06	.249	2.07	.165	1.06	.082	.53	.041
.8	6.38	4.31	.302	3.27	.297	2.21	.201	1.14	.104	.57	.052
.9	7.62	4.57	.467	3.48	.356	2.35	.240	1.20	.122	.60	.061
1.0	8.025	4.87	.554	3.67	.417	2.48	.281	1.26	.144	.63	.072
1.25	8.99	5.29	.751	4.02	.571	2.71	.385	1.39	.197	.69	.098
1.50	9.83	5.92	1.01	4.50	.767	3.03	.517	1.55	.259	.77	.129
1.75	10.59	6.40	1.27	4.86	.967	3.27	.650	1.67	.326	.83	.163
2.00	11.35	6.85	1.56	5.20	1.18	3.50	.795	1.79	.398	.89	.199
2.25	12.00	7.27	1.86	5.51	1.41	3.71	.049	1.89	.475	.95	.237
2.50	12.68	7.67	2.18	5.81	1.65	3.91	1.11	1.99	.565	1.00	.283
2.75	13.32	8.05	2.53	6.09	1.86	4.10	1.28	2.09	.654	1.04	.327
3.00	13.90	8.41	2.8	6.36	2.17	4.27	1.46	2.18	.743	1.09	.371
3.50	15.01	9.08	3.61	6.86	2.73	4.61	1.83	2.35	.935	1.17	.467
4.00	16.05	9.97	4.54	7.32	3.33	4.92	2.24	2.50	1.14	1.25	.568
4.50	17.02	10.29	5.26	7.75	3.96	5.21	2.66	2.65	1.36	1.32	.678
5.00	17.95	10.84	6.16	8.16	4.64	5.49	3.12	2.78	1.58	1.39	.781
6	19.66	11.84	8.08	8.91	6.08	5.98	4.08	3.03	2.07	1.51	1.03
7	21.23	12.76	10.14	9.61	7.64	6.43	5.12	3.24	2.58	1.62	1.29
8	22.71	13.64	12.40	10.25	9.32	6.84	6.22	3.45	3.14	1.71	1.50
9	24.70	14.47	14.80	10.86	11.11	7.25	7.42	3.64	3.72	1.83	1.82
10	25.38	15.25	17.34	11.44	13.00	7.62	8.66	3.83	4.34	1.92	2.18
15	31.08	18.68	31.85	14.01	23.88	9.34	15.93	4.60	8.00	2.36	4.02
20	35.89	21.50	49.05	16.18	36.78	10.8	24.55	5.42	12.29	2.72	6.15
25	40.13	24.12	68.52	18.10	51.42	12.08	34.32	6.06	17.22	3.05	8.67
30	43.95	26.43	90.10	19.84	67.64	13.47	45.92	6.64	22.64	3.35	11.42
35	47.47	28.55	113.6	21.44	85.27	14.31	56.06	7.18	28.56	3.62	14.40
40	50.75	30.53	138.8	22.94	104.3	15.32	69.64	7.68	34.91	3.79	17.23
45	53.83	32.39	165.6	24.35	124.5	16.26	83.14	8.16	41.73	4.12	21.02
50	56.75	34.15	194.0	25.68	145.8	17.16	97.50	8.61	48.92	4.35	24.72

VOLUMES OF WATER REQUIRED FOR EFFECTIVE USE IN OPERATING HYDRAULIC GIANTS.

Head. Feet.	2-Inch Nozzle. Miner's Inches.	2½-Inch Nozzle. Miner's Inches.	3-Inch Nozzle. Miner's Inches.	4-Inch Nozzle. Miner's Inches.	5-Inch Nozzle. Miner's Inches.
100	80	125	185	325	500
150	100	155	225	400	625
200	115	180	260	460	715
250	130	200	290	515	800
300	140	220	320	565	880
350	150	240	345	610	950
400	160	255	365	650	1000

The area of circles in square feet may be obtained from the following table, — which is also the number of cubic feet in 1 foot length of the pipe. (Trautwine.)

Diameter. Inches.	Area. Sq. Feet.	Diameter. Inches.	Area. Sq. Feet.	Diameter Inches.	Area. Sq. Feet.
$\frac{1}{4}$	.0003	$3\frac{1}{4}$	.0576	$6\frac{1}{4}$	.2131
$\frac{1}{2}$	.0014	$3\frac{1}{2}$	.0668	$6\frac{1}{2}$	.2304
$\frac{3}{4}$	.0031	$3\frac{3}{4}$	.0767	$6\frac{3}{4}$	.2485
1	.0055	4	.0873	7	.2673
$1\frac{1}{4}$	.0085	$4\frac{1}{4}$	.0985	$7\frac{1}{4}$	.2867
$1\frac{1}{2}$	.0123	$4\frac{1}{2}$	.1104	$7\frac{1}{2}$	.3068
$1\frac{3}{4}$	.0167	$4\frac{3}{4}$	.1231	$7\frac{3}{4}$	.3276
2	.0218	5	.1363	8	.3491
$2\frac{1}{4}$	.0276	$5\frac{1}{4}$	.1503	$8\frac{1}{4}$	.3712
$2\frac{1}{2}$	.0341	$5\frac{1}{2}$	.1650	$8\frac{1}{2}$	.3941
$2\frac{3}{4}$	.0412	$5\frac{3}{4}$	.1803	$8\frac{3}{4}$	.4176
3	.0491	6	.1964	9	.4418

To Find the Square Root of a Number: Separate the given number into periods of two each, beginning at unit's place, thus: 18, 66, 24; or if the number be a decimal fraction, work both right and left from unit's place, thus: 1, 96, 13, 69.

Find the greatest number whose square will go into the first period, and subtract this square; to the remainder annex the next period. Divide this new dividend by twice the square root already found, multiplied by 10 for a trial divisor. The quotient thus found add to the trial divisor; it is the next figure of the root. Multiply this divisor by the last root figure and subtract as in the first instance, etc.

Example. — Find the square root of 186624.

$$
\begin{array}{r}
18,\ 66,\ 24\ \overline{|432} \\
4 \times 4 = \quad 16 \\
4 \times 2 = 8 \times 10 = 80 + 3 = 83)\ 266 \\
83 \times 3 = \quad 249 \\
80 + 3 \times 2 = 86 \times 10 = 860 + 2 = 862\ \ 1724 \\
862 \times 2 = \quad 1724
\end{array}
$$

Example. — Find the square root of 58.140625.

$$
\begin{array}{r}
58.\ 14\ 06\ 25\ \overline{|7.625}\ Ans. \\
49 \\
7 \times 2 = 14 \times 10 = 140 + 6 = 146 \quad 914 \\
146 \times 6 = 876 \\
.3806
\end{array}
$$

$$
\begin{array}{l}
140 \\
\underline{12 = 6 \times 2} \\
152 \times 10 = 1520 + 2 = 1522 \\
\qquad 1522 \times 2 = 3044 \\
1520 \qquad\qquad\qquad 76225 \\
\underline{4 = 2 \times 2} \\
1524 \times 10 = 15240 + 5 = 15245 \\
\qquad 15245 \times 5 = 76225
\end{array}
$$

Example. — Find the square root of 196.1369.

$$\text{\textit{Answer.}}\quad 14.0048 + .$$

TABLE OF SAFE HEAD FOR RIVETED HYDRAULIC PIPE.

SHOWING PRICE AND WEIGHT WITH SAFE HEAD FOR VARIOUS SIZES OF DOUBLE-RIVETED PIPE.

Diameter of Pipe in Inches.	Area of Pipe in Inches.	Thickness of Iron by Wire Gauge.	Head in Feet the Pipe will safely stand.	Cub. Ft. of Water Pipe will convey per min. at Vel. 3 ft. per second.	Weight per Lineal Foot in Lbs.	Price per Foot.
3	7	18	400	9	2	$0.20
4	12	18	350	16	$2\frac{1}{4}$	.25
4	12	16	525	16	3	.35
5	20	18	325	25	$3\frac{1}{2}$	.35
5	20	16	500	25	$4\frac{1}{4}$	.45
5	20	14	675	25	5	.50
6	28	18	296	36	$4\frac{1}{4}$	.44
6	28	16	487	36	$5\frac{3}{4}$	.50
6	28	14	743	36	$7\frac{1}{2}$	.56
7	38	18	254	50	$5\frac{1}{4}$	.50
7	38	16	419	50	$6\frac{3}{4}$	.56
7	38	14	640	50	$8\frac{1}{2}$	.63
8	50	16	367	63	$7\frac{1}{2}$	.65
8	50	14	560	63	$9\frac{1}{2}$	.75
8	50	12	854	63	13	.94
9	63	16	327	80	$8\frac{1}{2}$	.69
9	63	14	499	80	$10\frac{3}{4}$	.88
9	63	12	761	80	$14\frac{1}{4}$	1.06
10	78	16	295	100	$9\frac{1}{4}$	.72
10	78	14	450	100	$11\frac{5}{8}$	.82
10	78	12	687	100	$15\frac{3}{4}$	1.00
10	78	11	754	100	$17\frac{1}{2}$	1.25
10	78	10	900	100	$19\frac{1}{4}$	1.50
11	95	16	269	120	$9\frac{3}{4}$	.75
11	95	14	412	120	13	.94
11	95	12	626	120	$17\frac{1}{2}$	1.25
11	95	11	687	120	$18\frac{3}{4}$	1.44
11	95	10	820	120	21	1.62
12	113	16	246	142	$11\frac{1}{4}$	.82
12	113	14	377	142	14	1.00
12	113	12	574	142	$18\frac{1}{2}$	1.38
12	113	11	630	142	$19\frac{3}{4}$	1.50
12	113	10	753	142	$22\frac{1}{4}$	1.69

SAFE HEAD FOR RIVETED HYDRAULIC PIPE.—(*Continued.*)

Diameter of Pipe in Inches.	Area of Pipe in Inches.	Thickness of Iron by Wire Gauge.	Head in Feet the Pipe will safely stand.	Cub. Ft. of Water Pipe will convey per min. at Vel. 3 ft. per second.	Weight per Lineal Foot in Lbs.	Price per Foot.
13	132	16	228	170	12	$0.90
13	132	14	348	170	15	1.12
13	132	12	530	170	20	1.50
13	132	11	583	170	22	1.65
13	132	10	696	170	24½	1.80
14	153	16	211	200	13	.98
14	153	14	324	200	16	1.17
14	153	12	494	200	21½	1.57
14	153	11	543	200	23½	1.72
14	153	10	648	200	26	1.95
15	176	16	197	225	13¾	.96
15	176	14	302	225	17	1.28
15	176	12	460	225	23	1.75
15	176	11	507	225	24½	1.95
15	176	10	606	225	28	2.10
16	201	16	185	255	14½	1.05
16	201	14	283	255	17¼	1.20
16	201	12	432	255	24¼	1.70
16	201	11	474	255	26½	1.85
16	201	10	567	255	29½	2.00
18	254	16	165	320	16½	1.20
18	254	14	252	320	20½	1.40
18	254	12	385	320	27¼	1.90
18	254	11	424	320	30	2.10
18	254	10	505	320	34	2.40
20	314	16	148	400	18	1.26
20	314	14	227	400	22½	1.54
20	314	12	346	400	30	2.10
20	314	11	380	400	32½	2.25
20	314	10	456	400	36½	2.50
22	380	16	135	480	20	1.40
22	380	14	206	480	24¾	1.70
22	380	12	316	480	32¾	2.25
22	380	11	347	480	35¾	2.45
22	380	10	415	480	40	2.80
24	452	14	188	570	27¼	1.80
24	452	12	290	570	35½	2.35
24	452	11	318	570	39	2.70
24	452	10	379	570	43½	2.95
24	452	8	466	570	53	3.50

SAFE HEAD FOR RIVETE DHYDRAULIC PIPE.—(*Continued.*)

Diameter of Pipe in Inches.	Area of Pipe in Inches.	Thickness of Iron by Wire Gauge.	Head in Feet the Pipe will safely stand.	Cub. Ft. of Water Pipe will convey per min. at Vel. 3 ft. per second.	Weight per Lineal Foot in Lbs.	Price per Foot.
26	530	14	175	670	29¼	$2.00
26	530	12	267	670	38½	2.59
26	530	11	294	670	42	2 87
26	530	10	352	670	47	3.10
26	530	8	432	670	57¼	3.85
28	615	14	102	775	31¼	2.12
28	615	12	247	775	41¼	2.75
28	615	11	273	775	45	3.00
28	615	10	327	775	50¼	3.20
28	615	8	400	775	61¼	4.15
30	706	12	231	890	44	2.90
30	706	11	254	890	48	3.15
30	706	10	304	890	54	3.50
30	706	8	375	890	65	4.30
30	706	7	425	890	74	4.75
36	1017	11	141	1300	58	3.80
36	1017	10	155	1300	67	4.30
36	1017	8	192	1300	78	5.10
36	1017	7	210	1300	88	5.75
40	1256	10	141	1600	71	4.75
40	1256	8	174	1600	86	5.60
40	1256	7	189	1600	97	6.40
40	1256	6	213	1600	108	7.35
40	1256	4	250	1600	126	8.50
42	1385	10	135	1760	74½	5.05
42	1385	8	165	1760	91	6.20
42	1385	7	180	1760	102	7.00
42	1385	6	210	1760	114	7.80
42	1385	4	240	1760	133	9.00
42	1385	1¼	270	1760	137	9.50
42	1385	3	300	1760	145	10.00
42	1385	5⁄16	321	1760	177	12.00
42	1385	⅜	363	1760	216	15.00

NOTE.—Where formed and punched including rivets, for mule packing or to facilitate transportation by other means, 30 per cent may be deducted from prices above given. This list is based upon pipe coated inside and out with asphaltum, and is given for the purpose of enabling parties to make an approximate estimate of the cost. Net prices will be quoted on application.

TABLE OF VELOCITIES.

Head in Feet.	Velocity, Feet per Second.	Actual Velocity, Feet per Second.	Discharge per Second through Nozzles.							
			1″	2″	3″	4″	5″	6″	7″	8″
10	25.4	26.32	11.18	44.30	99.78	177.4	277.0	399.1	682.2	709.4
20	35.9	28 72	15.79	62.61	141.0	250.8	391.4	564.1	767.7	1026
30	43.9	35.12	19.32	76.56	173.9	306.6	478.7	689.7	938.8	1226
40	50.7	40.56	22.31	89.24	199.3	354.1	552.8	796.7	1085	1416
50	56.7	45.36	24.95	98.88	222.7	396.0	618.4	890.9	1213	1584
60	62.1	49.68	27.33	108.30	243.9	433.7	677.1	975.7	1328	1735
70	67.1	53.68	29.52	117.01	263.5	468.6	731.7	1053	1435	1874
80	71.8	57.44	31.66	125.24	282.0	501.5	783.9	1129	1535	2005
90	76.1	60.88	33.49	132.87	298.9	531.4	829.8	1196	1627	2126
100	80.3	64.24	35.33	140.32	308.9	560.8	874.6	1262	1717	2243
110	84.2	67.36	37.05	146.82	330.8	588.0	918.1	1323	1801	2352
120	87.96	70.36	38.70	153.48	345.5	614.3	959.0	1382	1881	2456
130	91.54	73.23	40.28	159.92	359.5	639.3	998.1	1439	1958	2556
140	94.99	75.99	41.80	165.73	373.1	663.4	1036	1493	2031	2653
150	98.3	78.64	43.26	171.54	386.1	686.5	1072	1545	2102	2745
160	101.49	81.19	44.65	177.26	398.5	708.8	1107	1595	2170	2834
170	104.56	83.62	45.99	182.36	410.5	718.4	1140	1643	2236	2919
180	107.76	86.20	47.41	188.16	423.2	752.5	1176	1693	2305	3009
190	110.65	88.52	48.69	192.92	434.7	772.8	1207	1739	2366	3091
200	113.54	90.83	49.94	198.16	446.0	793.0	1238	1784	2428	3171
210	116.35	93.08	51.20	203.80	457.0	812.6	1259	1828	2488	3250
220	119.08	95.26	52.49	207.82	467.8	831.7	1299	1872	2547	3326
230	121.73	97.38	53.56	212.33	478.1	850.1	1323	1913	2604	3400
240	124	99.2	54.56	216.2	487.0	866.0	1352	1948	2652	3463
250	126	100.8	55.44	219.8	495.0	888.0	1374	1980	2694	3519
260	129	103.2	56.76	225.0	560.7	900.9	1407	2027	2759	3603
270	131	104.8	57.64	228.5	514.6	914.9	1428	2059	2801	3649
280	134	107.2	58 96	233.7	526.3	935.9	1461	2105	2865	3742
290	136	108.8	59.84	237.7	534.1	949.9	1483	2127	2909	3798
300	139	111.2	60.32	239.1	538.8	957.5	1495	2154	2932	3819
310	141	112.8	61.64	240.3	541.2	962.3	1503	2165	2947	3835
320	143	114.4	62.92	249.4	561.7	998.8	1559	2246	3058	3993
330	145	116.0	63.80	252.9	569.6	1012	1585	2278	3100	4050
340	148	118.4	64 61	256.1	576.8	1025	1601	2307	3140	4101
350	150	120.0	66.00	261.6	589.2	1047	1635	2357	3207	4189
360	152	121.6	66.88	262.5	597.1	1062	1658	2388	3245	4245
370	154	123.2	67.76	265.1	604.9	1075	1679	2419	3293	4301
380	156	124.8	68.64	272.1	612.8	1090	1701	2449	3336	4358
390	158	126.4	69.52	275.6	620.6	1104	1723	2482	3402	4412
400	160	128.0	70.40	279.0	628.5	1117	1746	2514	3422	4468
410	162	129.6	71.28	282.5	636.3	1132	1767	2545	3462	4523
420	164	131.2	72.11	285.8	643.7	1144	1787	2574	3505	4573
430	166	132.8	73.04	289.5	652.0	1160	1790	2608	3539	4586
440	168	134.4	73.92	293.0	659.9	1174	1832	2640	3593	4692
450	170	136.0	74.80	296.4	667.8	1188	1854	2672	3635	4747

TABLE FOR WEIR MEASUREMENT,

GIVING CUBIC FEET OF WATER PER MINUTE THAT WILL FLOW OVER A WEIR I INCH WIDE AND FROM $\frac{1}{8}$ TO $20\frac{7}{8}$ INCHES DEEP.

Inches.	⅛	¼	⅜	½	⅝	¾	⅞
0	.00	.01	.05	.09	.14	.19	.26	.32
1	.40	.47	.55	.64	.73	.82	.92	1.02
2	1.13	1.23	1.35	1.46	1.58	1.70	1.82	1.95
3	2.07	2.21	2.34	2.48	2.61	2.76	2.90	3.05
4	3.20	3.35	3.50	3.66	3.81	3.97	4.14	4.30
5	4.47	4.64	3.81	4.98	5.15	5.33	5.51	5.69
6	5.87	6.06	6.25	6.44	6.62	6.82	7.01	7.21
7	7.40	7.60	7.80	8.01	8.21	8.42	8.63	8.83
8	9.05	9.26	9.47	9.69	9.91	10.13	10.35	10.57
9	10.80	11.02	11.25	11.48	11.71	11.94	12.17	12.41
10	12.64	12.88	13.12	13.36	13.60	13.85	14.09	14.34
11	14.59	14.84	15.09	15.34	15.59	15 85	16.11	16.36
12	16.62	16.88	17.15	17.41	17.67	17.94	18.21	18.47
13	18.74	19.01	19.29	19.56	19.84	20.11	20.39	20.67
14	20.95	21.23	21.51	21.80	22.08	22.37	22.65	22.94
15	23.23	23.52	23.82	24.11	24.40	24.70	25.00	25.30
16	25.60	25.90	26.20	26.50	26.80	27.11	27.42	27.72
17	28.03	28.34	28.65	28.97	29.28	29.59	29.91	30.22
18	30.54	30.86	31.18	31.50	31.82	32.15	32.47	32.80
19	33.12	33.45	33.78	34.11	34.44	34.77	35.10	35.44
20	35.77	36.11	36.45	36.78	37.12	37.46	37.80	38.15

LOSS OF HEAD IN PIPE BY FRICTION.

The following tables show the loss of head by friction in each 100 feet in length of different diameters of pipe when discharging the following quantities of water per minute:

INSIDE DIAMETER OF PIPE IN INCHES.

Velocity in Feet per Sec.	1		2		3		4		5		6	
	Loss of Head in Feet.	Cubic Feet per Min.	Loss of Head in Feet.	Cubic Feet per Min.	Loss of Head in Feet.	Cubic Feet per Min.	Loss of Head in Feet.	Cubic Feet per Min.	Loss of Head in Feet.	Cubic Feet per Min.	Loss of Head in Feet.	Cubic Feet per Min.
2.0	2.37	.65	1.185	2.62	.791	5.89	.593	10.4	.474	16.3	.395	23.5
2.2	2.80	.73	1.404	2.88	.936	6.48	.702	11.5	.561	18	.468	25.9
2.4	3.27	.79	1.639	3.14	1.093	7.07	.819	12.5	.650	19.6	.547	28.2
2.6	3.78	.86	1.891	3.40	1.26	7.65	.945	13.6	.757	21.3	.631	30.6
2.8	4.32	.92	2.16	3.66	1.44	8.24	1.080	14.6	.864	22.9	.720	32.9
3.0	4.89	.99	2.44	3.92	1.62	8.83	1.22	15.7	.978	24.5	.815	35.3
3.2	5.47	1.06	2.73	4.18	1.82	9.42	1.37	16.7	1.098	26.2	.915	37.7
3.4	6.09	1.12	3.05	4.45	2.04	10.00	1.52	17.8	1.22	27.8	1.021	40
3.6	6.76	1.19	3.38	4.71	2.26	10.60	1.69	18.8	1.35	29.4	1.131	42.4
3.8	7.48	1.26	3.74	4.97	2.49	11.20	1.87	19.9	1.49	31	1.25	44.7
4.0	8.20	1.32	4.10	5.23	2.73	11.80	2.05	20.9	1.64	32.7	1.37	47.1
4.2	8.97	1.39	4.49	5.49	2.98	12.30	2.24	22.0	1.79	34.3	1.49	49.5
4.4	9.77	1.45	4.89	5.76	3.25	12.90	2.43	23.0	1.95	36.0	1.62	51.8
4.6	10.60	1.52	5.30	6.02	3.53	13.50	2.64	24.0	2.11	37.6	1.76	54.1
4.8	11.45	1.58	5.72	6.28	3.81	14.10	2.85	25.1	2.27	39.2	1.90	56.5
5.0	12.33	1.65	6.17	6.54	4.11	14.70	3.08	26.2	2.46	40.9	2.05	58.0
5.2	13.24	1.72	6.62	6.80	4.41	15.30	3.31	27.2	2.65	42.5	2.21	61.2
5.4	14.20	1.78	7.10	7.06	4.73	15.90	3.55	28.2	2.84	44.2	2.37	63.6
5.6	15.16	1.85	7.58	7.32	5.06	16.50	3.79	29.3	3.03	45.8	2.53	65.9
5.8	16.17	1.91	8.09	7.58	5.40	17.10	4.04	30.3	3.24	47.4	2.70	68.3
6.0	17.23	1.98	8.61	7.85	5.74	17.70	4.31	31.4	3.45	49.1	2.87	70.7
7.0	22.89	2.31	11.45	9.16	7.62	20.60	5.72	36.6	4.57	57.2	3.81	82.4

LOSS OF HEAD IN PIPE BY FRICTION—(Continued.)

Inside Diameter of Pipe in Inches.

Velocity in Feet per Sec.	7		8		9		10		11		12	
	Loss of Head in Feet.	Cubic Feet per Min.	Loss of Head in Feet.	Cubic Feet per Min.	Loss of Head in Feet.	Cubic Feet per Min.	Loss of Head in Feet.	Cubic Feet per Min.	Loss of Head in Feet.	Cubic Feet per Min.	Loss of Head in Feet.	Cubic Feet per Min.
2.0	.338	32.0	.296	41.9	.264	53	.237	65.4	.216	79.2	.198	94.2
2.2	.401	35.3	.351	46.1	.312	58.3	.281	72	.255	87.1	.234	103
2.4	.468	38.5	.410	50.2	.365	63.6	.327	78.5	.297	95.0	.273	113
2.6	.540	41.7	.473	54.4	.420	68.9	.378	85.1	.344	103	.315	122
2.8	.617	44.9	.540	58.6	.480	74.2	.432	91.6	.392	111	.360	132
3.0	.698	48.1	.611	62.8	.544	79.5	.488	98.2	.444	119	.407	141
3.2	.785	51.3	.686	67	.609	84.8	.549	105	.499	127	.457	151
3.4	.875	54.5	.705	71.2	.680	90.1	.612	111	.557	134	.510	160
3.6	.969	57.7	.848	75.4	.755	95.4	.679	118	.617	142	.566	169
3.8	1.070	60.9	.936	79.6	.831	101	.749	124	.680	150	.624	179
4.0	1.175	64.1	1.027	83.7	.913	106	.822	131	.747	158	.685	188
4.2	1.28	67.3	1.122	87.9	.998	111	.897	137	.816	166	.749	198
4.4	1.39	70.5	1.22	92.1	1.086	116	.977	144	.888	174	.815	207
4.6	1.51	73.7	1.32	96.3	1.177	122	1.059	150	.963	182	.883	217
4.8	1.63	76.9	1.43	100.0	1.27	127	1.145	157	1.040	190	.954	226
5.0	1.76	80.2	1.54	105	1.37	132	1.23	163	1.122	198	1.028	235
5.2	1.89	83.3	1.65	109	1.47	138	1.32	170	1.20	206	1.104	245
5.4	2.03	86.6	1.77	113	1.57	143	1.41	177	1.28	214	1.183	254
5.6	2.17	89.8	1.89	117	1.68	148	1.51	183	1.37	222	1.26	264
5.8	2.31	93.0	2.01	121	1.80	154	1.61	190	1.46	229	1.34	273
6.0	2.46	96.2	2.15	125	1.92	159	1.71	196	1.56	237	1.43	283
7.0	3.26	112.0	2.85	146	2.52	185	2.28	229	2.07	277	1.91	330

LOSS OF HEAD IN PIPE BY FRICTION—(*Continued.*)

INSIDE DIAMETER OF PIPE IN INCHES.

Velocity in Feet per Sec.	13		14		15		16		18		20	
	Loss of Head in Feet.	Cubic Feet per Min.	Loss of Head in Feet.	Cubic Feet per Min.	Loss of Head in Feet.	Cubic Feet per Min.	Loss of Head in Feet.	Cubic Feet per Min.	Loss of Head in Feet.	Cubic Feet per Min.	Loss of Head in Feet.	Cubic Feet per Min.
2.0	.183	110	.169	128	.158	147	.147	167	.132	212	.119	262
2.2	.216	121	.200	141	.187	162	.175	184	.156	233	.140	288
2.4	.252	133	.234	154	.218	176	.205	201	.182	254	.164	314
2.6	.290	144	.270	167	.252	191	.236	218	.210	275	.189	340
2.8	.332	156	.308	179	.288	206	.270	234	.240	297	.216	366
3.0	.375	166	.349	192	.325	221	.306	251	.271	318	.245	393
3.2	.422	177	.392	205	.366	235	.343	268	.305	339	.275	419
3.4	.471	188	.438	218	.408	250	.383	284	.339	360	.306	445
3.6	.522	199	.485	231	.452	265	.425	301	.377	382	.339	471
3.8	.576	210	.535	243	.499	280	.468	318	.416	403	.374	497
4.0	.632	221	.587	256	.548	294	.513	335	.456	424	.410	523
4.2	.691	232	.641	269	.598	309	.561	352	.499	445	.449	550
4.4	.751	243	.698	282	.651	324	.611	368	.542	466	.488	576
4.6	.815	254	.757	295	.707	339	.662	385	.588	488	.529	602
4.8	.881	265	.818	308	.763	353	.715	402	.636	509	.572	628
5.0	.949	276	.881	321	.822	368	.770	419	.685	530	.617	654
5.2	1.020	287	.947	333	.883	383	.828	435	.730	551	.662	680
5.4	1.092	298	1.014	346	.947	397	.888	452	.788	572	.710	707
5.6	1.167	309	1.083	359	1.011	412	.949	469	.843	594	.758	733
5.8	1.245	321	1.155	372	1.078	427	1.011	486	.899	615	.809	759
6.0	1.325	332	1.229	385	1.148	442	1.076	502	.957	636	.861	785
7.0	1.75	387	1.63	449	1.52	515	1.43	586	1.27	742	1.143	916

LOSS OF HEAD IN PIPE BY FRICTION—(Continued.)

INSIDE DIAMETER OF PIPE IN INCHES.

Velocity in Feet per Sec.	22		24		26		28		30		56	
	Loss of Head in Feet.	Cubic Feet per Min.	Loss of Head in Feet.	Cubic Feet per Min.	Loss of Head in Feet.	Cubic Feet per Min.	Loss of Head in Feet.	Cubic Feet per Min.	Loss of Head in Feet.	Cubic Feet per Min.	Loss of Head in Feet.	Cubic Feet per Min.
2.0	.108	316	.098	377	.091	442	.084	513	.079	589	.066	848
2.2	.127	348	.116	414	.108	486	.099	564	.093	648	.078	933
2.4	.149	380	.136	452	.126	531	.116	616	.109	707	.091	1018
2.6	.171	412	.157	490	.145	575	.134	667	.126	766	.104	1100
2.8	.195	443	.180	528	.165	619	.153	718	.144	824	.119	1188
3.0	.222	475	.204	565	.188	663	.174	770	.163	883	.135	1273
3.2	.249	507	.229	603	.211	708	.195	821	.182	942	.152	1357
3.4	.278	538	.255	641	.235	752	.218	872	.204	1001	.169	1442
3.6	.308	570	.283	678	.261	796	.242	923	.226	1060	.188	1527
3.8	.340	601	.312	716	.288	840	.267	974	.249	1119	.207	1612
4.0	.373	633	.342	754	.315	885	.293	1026	.273	1178	.228	1697
4.2	.408	665	.374	791	.345	929	.320	1077	.299	1237	.249	1782
4.4	.444	697	.407	829	.375	973	.348	1129	.325	1296	.271	1866
4.6	.482	728	.441	867	.407	1017	.378	1180	.353	1355	.294	1951
4.8	.521	760	.476	905	.440	1062	.409	1231	.381	1414	.318	2036
5.0	.561	792	.513	942	.474	1106	.440	1283	.411	1472	.312	2121
5.2	.602	823	.552	980	.510	1150	.473	1334	.441	1531	.368	2206
5.4	.645	855	.591	1018	.546	1194	.507	1385	.473	1590	.394	2291
5.6	.690	887	.632	1055	.583	1239	.542	1437	.506	1649	.421	2376
5.8	.735	918	.674	1093	.622	1283	.578	1488	.540	1708	.450	2460
6.0	.782	950	.717	1131	.662	1327	.615	1539	.574	1767	.479	2545
7.0	1.040	1109	.953	1319	.879	1548	.817	1796	.762	2061	.636	2868

Example.—Have 200 feet head and 600 feet of 11-inch pipe, carrying 119 cubic feet of water per minute. To find effective head: In right-hand column under 11-inch pipe find 119 cubic feet; opposite this will be found the coefficient of friction for this amount of water, which is .444. Multiply this by the number of hundred feet of pipe, which is 6, and you will have 2.66 feet, which is the loss of head. Therefore the effective head is 200—2.66 = 197.34.

INDEX

345

FINIS